EUL VERLAG

Richard Vahrenkamp

The German Autobahn 1920–1945

Hafraba Visions and Mega Projects

EUL VERLAG

Bibliografische Information der Deutschen Nationalbibliothek

Die Deutsche Nationalbibliothek verzeichnet diese Publikation in der Deutschen Nationalbibliografie; detaillierte bibliografische Daten sind im Internet über <http://dnb.d-nb.de> abrufbar.

ISBN 978-3-89936-940-3
1. Auflage Juli 2010

Coverbild: Opening of the Autobahn Cologne–Bonn on 6 August 1932, source: Municipial Archive Bonn, image no. 4738.

JOSEF EUL VERLAG GmbH
Brandsberg 6
53797 Lohmar
Tel.: 0 22 05 / 90 10 6-6
Fax: 0 22 05 / 90 10 6-88
E-Mail: info@eul-verlag.de
http://www.eul-verlag.de

Bei der Herstellung unserer Bücher möchten wir die Umwelt schonen. Dieses Buch ist daher auf säurefreiem, 100% chlorfrei gebleichtem, alterungsbeständigem Papier nach DIN 6738 gedruckt.

Table of Contents

Images

Tables

1 Autobahn and the Theory of Infrastructure

Transport infrastructures have a peculiar interim status. Once in place, they are quickly taken for granted but the lead up to the implementation process is time-consuming, capital intensive and risky. Grand visions are required to mobilize the necessary political and economic forces. And it is not the completed infrastructure systems per se that are of interest to the public but rather the possibilities that they open for private and business interests – as well as for the public sector. What remains to be dealt with now are issues concerning access, maintenance, financing and increasing capacity as well as the vulnerability and path dependency of infrastructures.[1]

The expressway network in Europe developed into an essential infrastructure of the 20th and the beginning of the 21st century that provided a means of commuting, as well as accommodated leisure travel and the cargo supply for a mass consumption society. This book discusses how expressways were developed in Europe in the 1920s and 1930s. It focuses on the various forerunner projects and the role of the Hafraba association which was significant in the Hessian region, with its actors in Kassel, Frankfurt and Darmstadt. It is shown how the *autobahn* concept developed from the Italian expressways and the Bonn–Cologne *autobahn* to the design of the Nazi *autobahn* project. The Bonn–Cologne *autobahn* – Germany's first *autobahn* – was opened in 1932 by Konrad Adenauer who would later become Chancellor of West-Germany. This *autobahn* section is explored here for the first time. The various regional legs that comprised the Nazi mega project: Frankfurt–Kassel–Göttingen, Frankfurt–Darmstadt–Heidelberg–Karlsruhe and the Munich–Salzburg are explored and subjected to scholarly attention for the first time. Moreover, the book shows how traffic on the *autobahn* developed and how driving on the *autobahn* was experienced.

With regard to the autobahn construction as a mega project, the transport policies of the National Socialists were highly contradictory. On the one hand, a network of 3000 km of high-speed roads was created in the years before 1939. But at the same time truck transport by

---

1 For the history of infrastructure see van Laak 2001, van Laak 2004, Vleuten and Kaijser 2006, Flyvbjerg, Bruzelius and Rothengatter 2003 and Shannon and Smets 2010.

private companies was drastically curtailed.[2] Indeed, the strategies of the NS transport policy do not even match its goal of rearmament. Neither were there incentives to promote private trucking enterprises, so that the army would have trucks to requisition in wartime, nor would the autobahn serve the deployment of troops.

This book shows how the autobahn construction in Hesse was subjected to irrational transport policy criteria with many arbitrary features. The aspects of landscape aesthetics, for example, took priority over transport engineering. The planning process also reveals the dictatorial role played by Fritz Todt, General Inspector of German Roads, and his unscrupulous exploitation of the goodwill of the Hafraba experts. And here too, in the field of road construction, we can see how shamelessly the Nazis adopted ideas from the Weimar Republic.

The a*utobahn* mega project started during the Nazi rule and maintained its momentum across the deep historical caesura of the year 1945. It is intriguing that in the early 1950s the Federal Republic of Germany implemented the Frankfurt Intersection "prestige project" exactly as planned in the NS era, even though its financial resources were acutely limited and traffic on the *autobahn* was sparse. This can count as evidence for the continuity of large-scale projects in completely different historical conditions.

In the past, research on Nazi *autobahn* construction from 1933–1939 has focused on topics such as the background, the decision-making process, the landscaping of the *autobahn*, the organization and financing of *autobahn* construction as well as the project's importance for job creation and the military.[3] Already in 1995, the wealth of studies prompted historian Ludwig to suggest that this field of research would become overcrowded.[4] However, a number of questions remain unanswered. The specific routing of the roads in certain parts of the country, for example, has not been examined nor has the work of the *autobahn* planning company Gezuvor. The road-rail rivalry also must be examined for its role in the decision-

---

2 See also Oswald Lutz: 3000 Kilometer Reichsautobahn für den gewerblichen Kraftverkehr, in: Die Strasse, 1938, issue 24, p. 772.

3 See appendix 2.

4 Karl-Heinz Ludwig: Politische Lösungen für technische Innovationen 1933–1945 – Eine antitechnische Mobilisierung, Ausformung und Instrumentalisierung der Technik, in: Technikgeschichte, Bd. 62, 1995, p. 333–344.

making processes. Moreover, the Chancellery files of the Hitler government that were published in recent years have yet to be evaluated. In particular, this study will focus on the routing of certain *autobahn* lines taking into account the relationship between local and long-distance traffic and the role of tourism in the general decision-making process for *autobahn* construction.

It has not been ascertained yet how the evolution of the a*utobahn* concept relates to the greatly increased motorization levels of the 1920s. The various studies on motorization in Germany during the 1920s, such as those by Edelmann and Flik, deal more with the taxation of vehicles and the automobile industry.[5] Moreover, they point out rationalization potentials by concentrating production and adopting American methods of mass production. They do not, however, link their approach to the politics of road construction. Only Mom, in comparing selected European countries, shows the correlation between motorization levels and the expressway construction policies.[6] Still, the question of the extent to which drivers used the a*utobahn* system in Germany before 1945 cannot be easily answered due to the fact that the level of motorization was very weak by European standards. Motorization rates remained weak in spite of support between 1933 and 1939 by the Nazi regime. It was not until the 1950s and 1960s that one could speak of mass motorization in Germany.

The question of what exactly is meant by the term "a*utobahn"* was posed systematically for the first time by Gabriel.[7] We will conform to his analysis and define "*autobahn*" in this paper as a junction-free road that is reserved for traffic by motor vehicles. Additional characteristics that have gradually developed over time are as follows: (1) The lanes going in one direction are spatially separated from the lanes going in the opposite direction; (2) this separation is either a median strip or a crash barrier; (3) there are at least two lanes in each direction; and (4) there is only limited access to the road in intervals of several kilometres. This definition can be further specified by the kinds of vehicles admissible. Motorcycles were initially

---

5 Reiner Flik: Von Ford lernen? Automobilbau und Motorisierung in Deutschland bis 1933, Cologne 2001.

6 Gijs Mom: Roads as Rails – European Highway-Network Building and the Desire for Long-Range Motorized Mobility, in: Technology & Culture, no. 4, vol. 46 (2005), pp. 745–772.

7 Gabriel, Roland: Deutsche Autobahnvorläufer – AVUS, Hafraba, München–Leipzig–Berlin, Leipzig–Halle, Düsseldorf–Köln–Bonn, in: Wolfgang Wirth: Die Autobahn – Von der Idee zur Wirklichkeit, Cologne 2005.

banned from the Bonn–Cologne *autobahn*, as well as from the expressways of northern Italy. On the park highways of Long Island, commercial verhicles were barred due to low-crossing bridges. In addition, the Genoa–Serravalle expressway was reserved for trucks only ("Camionale").[8] Long-distance expressways must also be distinguished from city expressways. In the book we use throughout the terms "autobahn" and "expressway". In the various countries of the world there are many other denotations: highway, motorway, autostrada, autoroute…

Antoine Picon has called for the integration of the history of planning and construction of expressways into a general theory of infrastructure.[9] Airports, container harbors, shopping centers and office blocks have the same appearance as expressways – they are facilities without a specific location ("non-locations"). Herewith, he refers to the differentiation of a generic space versus a specific space in architectural theory.[10] He describes the expressway as being generic because it endlessly reproduces the space of the lane. But at the same time, he describes the expressway as being specific because of its adaptation to the various national and environmental conditions.

In addition to the backward idealization, the *autobahn* network has developed at the same time a path-breaking function for the 20th and 21st century – and not only for mass motorization. During the 1950s, US President Dwight D. Eisenhower promoted an expressway project for the United States which was analogous to the German *autobahn*. On 22 February 1955, he delivered a message before Congress and proposed a nation-spanning highway network based on the German model. He justified his proposal with four main arguments:

1) Increased driving safety compared to country roads.
2) The economically desirable reduction of road transport costs.
3) An opportunity to evacuate large populations from the city in case of nuclear attack.

---

8 B. Joerges: Die Brücken des Robert Moses: Stille Post in der Stadt- und Techniksoziologie, in: Leviathan, 1999, issue 1. And: Lando Bortolotti: Origine e Primordi della Rete Autostradale in Italia 1922–1933, in: Storia Urbana, Band 59, 1992, p. 57. Rubinstein 2001, p. 164.

9 Antoine Picon: Towards a history of Infrastructure, in: Michele Bonino and Massimo Moraglio: Inventing Movement – History and Images of the A6 Motorway, Turin 2006, p. 13–17.

10 Jean Attali: The Roman System, or the Generic in All Times and Tenses, in: Rem Koolhaas (Editor): Mutations, Barcelona 2000.

4) The necessary adaptation of road networks to the rapid increase in motorization levels that were projected to continue in the future.[11]

In pushing his plan, he argued that expressways would offer a decentralized escape route for the population of big cities in the event of a nuclear war. But this argument was only supplementary to ensure the implementation of the project. In 1956, with the passing of the Federal Aid Highway Act, the plan could be realized.[12] In the 1970s, the US Pentagon viewed highways as a model for a fail-safe communication network for the Pentagon, whereby data packets could autonomously find their way through the network like individual motor vehicles. The network could be safeguarded through its manifold intersections. This basic model for a safe communication network combined with the by then-developed data transport protocol TCP/IP, was the basis for the development of the Internet infrastructure. The integration of the personal computer as the "vehicle of the spirit" into the Internet led to the "Internet boom" of the 1990s. In the 21st century, the Internet continues to influence communication in both business and private life.

This book is based on the evaluation of trade journals and national statistics. Various archives on the federal and state level were also inspected and the municipal archives along the Hafraba line from Hannover to Karlsruhe were reviewed. Additionally, the archives of the Chambers of Industry and Commerce in Leipzig, Cologne, Darmstadt and Munich were evaluated and the German Federal Archive's collection of press releases on questions of transport policy by the German Commercial Clerks Association were appraised. Edelmann's elaborate analyses of the political debates on transport issues between 1920 and 1939 were used as well.[13]

---

11 The message can be drawn from the server of the Eisenhower archive: http://eisenhower.archives.gov/Research/Digital_Documents/InterstateHighways/InterstateHighwaysdocuments.html.

12 Mark M. Rose: Interstate – Express Highway Politics 1939–1989, Tennessee UP, 1990, 2. improved edition, chapter 6. See also: Phil Patton: Open Road – A Celebration of the American Highway, New York 1986, p. 89 and Bruce E. Seely: Building the American Highway System–Engineers as Policy Makers, Philadelphia 1978, p. 213.

13 Heidrun Edelmann: Vom Luxusgut zum Gebrauchsgegenstand, Frankfurt 1989.

2 *Autobahn* Projects in Europe

In the Interbellum, various visions for expressway connections in Europe – and for entire networks – were the subject of discussions which have been evaluated in a recent study by Frank Schipper.[14] He underpins the leading role of Albert Thomas, the head of the International Labor Organization in Geneva – a branch of the League of Nations. Thomas proposed a series of infrastructure projects for Europe – power grids, inland navigation, railway networks and road networks – to provide work for the large number of unemployed and to serve as an impetus for the unification of Europe. In his role as the main speaker at the first International Congress on Expressways in Geneva from 31 August to 2 September 1931, he rallied support for this cause. Expressway proponents from Italy, France, United Kingdom, Belgium, the Netherlands, Switzerland and Germany had gathered for this event. A proposal for a network encompassing 14 000 km was made by the Technical Commission, chaired by Pietro Puricelli (cf. Image 1). On 18–20 April 1932 the second International Congress on Expressways took place in Milan and offered the opportunity for excursions to the newly-built expressway in Northern Italy.

In the Interbellum, most of the expressway projects outside Italy were mere proposals that generally didn't move beyond discussions. Only the Austrian project, the *Grossglockner* Alpine Pass, went into the implementation phase. At international road conferences, the long-distance routes London–Paris and Vienna–Istanbul were widely discussed.[15] Indeed, entire expressway networks spanning the whole of Europe were envisioned and pioneered by Puricelli, among others.[16] Road construction projects were also discussed in Austria and Switzerland. The *Schweizerische Autostrassen-Verein*, founded in 1929, debated projects with a focus on efficient road connections in Switzerland; the Bern–Thun road, which was supported by Marcel Nyffeler, was one such example.[17] In 1930, Austria started work on the *Grossglockner* Alpine Pass from Salzburg to Carinthia – a project that had already been under

---

[14] Schipper (2008), pp. 92–115.

[15] Die Autobahn 1 (1932), p. 1.

[16] Piero Puricelli (1934), p. 42. Puricelli presented his concept on the 7th International Road Conference in Munich, 3–8 September 1934.

[17] Nyfeller (1930), p. 6.

discussion in the 1920s.[18] Surprisingly, it had many parallels to the German *autobahn* project which began three years later. It was an integral part of a publicly debated job creation measure which it promoted as a national myth and exploited for domestic politics. The putting of this major project into the service of politics might well have served as the model that was later duplicated in the road construction policies of the Nazis in 1933.[19] The same mix of job creation, public and international attention, modernization and national myth can be seen in a giant project in the Soviet Union. That the building of the Belomor Canal from Leningrad to the White Sea in 1930 could be realized only through forced labor at the expense of a high mortality rate for the workers, did not dampen the international admiration. Indeed, Stalin elevated the project to the level of a national myth.[20]

Image 1: European Autobahn Network of Bureau International d'Autoroutes[21]

[18] Rigele (1998).

[19] Todt's journal Die Strasse reported on the Grossglockner alpine road for the first time in May 1935, see Örley (1935).

[20] Schlögel (2005).

[21] Heckmann-Strohkark, Ingrid: Der Traum von einer europäischen Gemeinschaft. Die internationalen Autobahnkongresse 1931 und 1932, in: Martin Heller and Andreas Volk (eds.): Die Schweizer Autobahn, Zürich, 1999,p. 40.

A Paris to Lille expressway was discussed in France.[22] Piero Puricelli, one of Europe's main expressway network proponents, fostered the idea at the 7th International Congress on Road Construction in Munich in 1934. Image 8 shows what he had in mind.[23] Many proposals were submitted for expressways between Italy and Germany and, in the 1920s and 1930s, Puricelli suggested a Berlin–Munich–Rome expressway.[24] On 4 September 1937, the *Völkische Beobachter* (The Folk Observer) published a blueprint for the alpine pass of this proposed expressway – it had previously already reported on the meetings conducted by Hitler, Puricelli and Todt focusing on the same project.[25]

---

22 *Hafraba* Mitteilungsblatt 3 (1929), p. 5.

23 Puricelli (1934), p. 42.

24 Bortolotti (1997).

25 Völkischer Beobachter (The Folk Observer) on 3 December 1936, Bayerisches Staatsarchiv (Bavarian State Archive), München (Munich), Autobahndirektion Südbayern, file 0368, and file 0405.

3 Constructing Expressways in Italy

German debates were inspired by the shining example of entrepreneur Piero Puricelli's expressway initiative in Milan. In Germany, 27 publications featuring the Italian expressways appeared from 1925 to 1939, as Hafen has revealed.[26] This section of the book looks at the peculiar social and economic factors that led to expressway construction in Italy.[27]

At first glance, the situation in Italy was similar to that in Germany. Both countries had a low per capita car density, whereas Italy's car density was consistently lower by half. However, this comparison of numbers does not take into account the varying regional distribution of automobiles. The industrial centers of Northern Italy had a per capita car density on par with the German cities. Thus, in the city of Turin, there was one car per 56 inhabitants in 1930 and in Milan this figure was one car per 60 inhabitants. These densities are comparable with (and even higher than) one car per 68 inhabitants in Frankfurt a.M. in 1930 (cf. Table 2).[28] Therefore, the cities in Northern Italy and Frankfurt had similar motivations to promote expressway projects.

As a consequence of warfare production in World War I, Northern Italy experienced rapid industrialization. The proliferation of FIAT production sites in Turin is evidence of this.[29] At the same time, a self-confident class of entrepreneurs with abundant capital became more visible in Northern Italy. It is not surprising that agitation against war-profiteers characterized the political arena during the economic crisis of 1920, and was also politically exploited by Mussolini.[30] It is still not clear whether reparation payments from Austria and Germany contributed to the increasing capital supply in Italy. The industrial progress in Northern Italy

---

26 Hafen (1956), pp. 30–32.

27 Zenone (2002).
Bortolotti (1992).
Moraglio (2002).

28 I owe the data from Turin and Milan to my colleague Massimo Moraglio, University of Turin. His sources are: Istat, Annuario statistico, Rome, Istat, 1930, and Città di Torino, Bollettino statistico, 1930.

29 Delle Donne (1995), p. 47.
See also Zenone (2002).

30 Rauscher (2001).

was linked with heavy automotive traffic in the cities and their suburbs and clearly stirred the need for expressways. The connection between expressway construction and the wave of Futurism – with its connotations of modernity and speed – is discussed by Daniela Zenone (2002).

In distinct contrast to the Hafraba project and to the Nazi *autobahn* project, the Italian initiatives did not have the goal of an *autobahn* network but were, without exception, decentralized. Individual communities, with the help of the local chambers of commerce and local leading businesses, seized the initiative. The projects were signed off by the government in Rome and approved by the Department of Public Works. Not all projects were completely privately financed. Some received a partial warranty from the state for deficiency suretyships or warranty interest, or otherwise received a partial grant to cover one third of the expenses – as happened for the Turin–Milan expressway owned by FIAT.[31] As Bergamo was cut off from railroad traffic the construction of the Trevilio–Rovato route demanded an expressway from Milan to Bergamo.

Route	Length in km	Company	Capital	Municipal investment in the construction costs in Lira	Permission with royal decree	Start of construction	Opening
Milan–Lakes	84.5	SA Autostrada	50 mio.		1922-12-17	1923-03-26	1924-09-21
Milan–Bergamo	48.8	SA Bergamasca	20 mio.	7 mio.	1925-09-04		1927-10-28
Naples–Pompeii	20	SA Autostrada Meridionali	15 mio.	Fixed interest by the state	1925-09-04		1928-10-28
Bergamo–Brescia	45	SA Bresicana with investment from the board of trade and municipal statutory corporations			1929-02-14		1905-04-14
Turin–Milan	125.8	owned by FIAT	30 mio.	3.7 mio. grant by the state per year	February 1930 decree-law	1930	1932-10-14
Florence–Viareggio	81	SA L'autostrada di Venecia		1/3 of the costs are borne by Venice and Padua, the remain by a 50-year government bond		1929-11-23	1933-10-15

[31] Hafraba Mitteilungsblatt 3 (1929), p. 5.

Route	Length in km	Company	Capital	Municipal investment in the construction costs in Lira	Permission with royal decree	Start of construction	Opening
Rome–Ostia	20			100% by the City of Rome			1928-10-28
Genoa–Serravalle	50			AASS			1935

Table 1: Sections of the Italian expressways
Source: Own research

Table 1 shows the individual projects which were, with two exceptions, founded as co-operations. Only the Rome–Ostia expressway belonged to the city of Rome. The Genoa–Seravalle expressway is a special case because it was reserved exclusively for truck traffic and was built for military purposes. It is notable that accessibility to tourist destinations for the residents of large cities in Italy influenced expressway construction projects: Milan–Lakes, Florence–Viareggio and Rome–Ostia.

The expressway promoter Piero Puricelli was a building contractor in Milan whose prominence led to an appointment as manager of the Milan Exhibition. Not surprisingly, Puricelli was able to mobilize private capital for his expressway project near the northern Italian lakes: By 1918, he had already secured financing for a research institute for road construction for the Italian Automobile Association TCI. Later private businesses put up the funds which were eventually also forthcoming from public organizations. In 1922, he presented a plan to the TCI for an expressway from Milan to the northern Italian lakes. Puricelli discussed the plan with Mussolini shortly after the latter seized power on 28 October 1922. Puricelli received approval from the government which was installed on 17 December 1922. Construction started in June 1923.[32]

[32] Bortolotti (1992), p. 46.

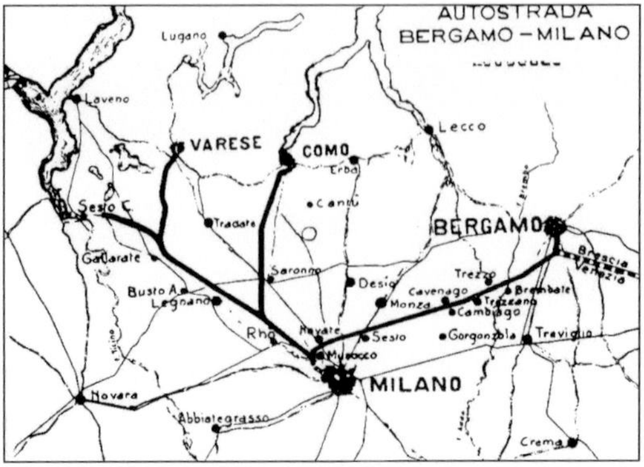

Image 2: Autobahn Network in North Italy in 1927[33]

The first route, Milan–Varese – shown in Image 2 – was conceived to shadow the railway. Just as a ticket was necessary to travel by rail, the operator of a car would be obliged to purchase a ticket at the expressway entrance. These tolls would essentially cover the construction costs. The tolls were staggered: For a medium-sized car of 17 to 26 HP a one-way ticket was (converted) 3.80 Marks.[34] The Milan–Varese route was 49 km long and opened to traffic on 20 September 1924. On 27 June 1925, the inauguration of the 36 km-long Milan–Como route followed. Both routes sported a road width of 11 meters on their common main route The main route exhibited three lanes with the middle one reserved for overtaking (cf. Image 3). The side route stretched along a length of 52 km and had a road width of 8m across two lanes.[35] Due to a low incidence of traffic, the approach with two or three lanes was sufficient to ensure fast car speeds. A speed limit of 40 km/h applied to trucks only.[36] At the time, these expressways constituted the first roads ever in Europe to be junction-free. And therefore, bridge constructions played a major role in the building of these roads. By the time that the roads had been extended to four lanes, however, the bridges proved to be a bottleneck for the flow of traffic and were also criticized for their appearance of functionality and lack of

---

33 Robert Otzen: Die Autostrasse Hansestädte–Frankfurt–Basel, Ziele und Zweck des Vereins „Hafraba", Hannover 1927, p. 14.

34 Uhlfelder (1930).

35 1. Sitzung des Technischen Ausschusses (1927), p. 9.

36 Uhlfelder (1930) reported on a daily average of 1,497 vehicles for the year 1928.

aesthetics (compare Image 12). The route was laid out with a layer of concrete – the material most suitable to withstand automobile traffic according to the American experience.[37]

In contrast to the Nazi *autobahn* project which did not have a strong military-political relevance, the development of Italy's Southern Alps chain for motorized traffic expressly incorporated defense-enhancing features. This effort was borne out of World War I experiences and explains the backing that this project received in Italy compared to the situation in Germany. However, this military-political interpretation is not supported by either Bortolotti's or by Moraglio's research.[38] There was, indeed, real misery that was experienced in Italy during the war. The unpaved roads had been wholly unsuitable for the heavy trucks that had supplied the front lines with war material and now efforts were underway to develop a new concept for the construction of durable roads made of concrete or asphalt. These, then, would, evolve into the expressways.

Image 3: Autobahn Mailand–Gallarate with three lanes[39]

To mark the anniversary of Mussolini's march into Rome on 28 October 1922, the expressway sections just mentioned were inaugurated with a propaganda spectacle. This linkage of

---

37 See also the discussion on concrete roads in the USA in the journal Verkehrstechnik, 1926 to 1930. In 1910, the first concrete roads were already being built in California, in order to create an alternative to transportation with the local railroad monopolies. See Gregory Thompson, The Passenger Train in the Motor Age: California Rail and Bus Industries 1910–1941, Ohio State Univ. Press, 1993.

38 Bortolotti (1992).

39 Lando Bortolotti and Guiseppe De Luca: Fascismo e autostrada, Mailand 1994, p. 54.

the expressway opening with politics might well have served as an example for the Nazi regime. Mussolini may have sought political benefits from this spectacle insofar as he was able to exploit it to cloak previous failures in Italy's road construction politics. The road network was in extremely poor condition in the 1930s. Bortolotti has pointed out that at the end of 1932, after 10 years of fascist control, only 3.2 per cent of the altogether 20 000 kilometres of country roads had either a solid or semi-solid surface.[40] Expressway construction, thus, was a symbolic policy of modernization but did not have a tangible impact for broad sections of the country. Obviously, the Department for Public Works was too weak to enforce a centralized plan for an expressway network. Instead, several entrepreneurs who were brimming with self-confidence, took on the construction of important routes themselves. Therefore, in Italy only unconnected segments were created rather than an entire network.

40 Ibidem, p. 45.

4 Motorization Levels in Germany from 1920 to 1930

The deficits of the German Reich's road system became obvious when it could no longer bear the increase in motor vehicle traffic. The number of registered motor cars in Germany rose quickly, tripling from 161 000 in 1925 to 433 000 by 1929.[41] The motor car steadily gained importance for business life in the 1920s, compelling the central association of German wholesale traders to survey its members on the frequency of their business-related car travel.[42] Table 2 below demonstrates the nearly two-fold increase in cars in the city of Frankfurt (Main) from 1927 to 1932.[43]

Year	Cars and buses	Cars and buses per 10,000 inhabitants	Number of inhabitants per Car and bus
1927	4278	90.2	110,8
1928	5955	108.2	92,4
1929	7284	131.8	75,8
1930	8134	147.9	67,6
1932	7912	148.0	67,6

Table 2: Development of Cars and Buses in Frankfurt (Main), 1927 to 1932

A significant increase in registered trucks can also be seen. The number of trucks rose in the 1920s at a significant average growth rate of 22 per cent annually according to the Berlin Institute of Research on Business Cycles.[44] The efficiency of transporting goods by truck in the 1920s rose simultaneously so that trucks gained advantages over transportation by train in terms of speed and price. According to research by Mellerowicz, the cost of truck transportation in the 1920s dropped 60 per cent due to better truck design and construction.[45] The number of small trucks increased most rapidly owing to their distinct advantage of being able

41 Statistisches Jahrbuch für das Deutsche Reich 1926, p. 103; 1930, p. 158. For the situation of the German car production see Bönig, pp. 421–439.

42 Stadtarchiv (City Archive) Frankfurt/Main, Akte IHK 854.

43 Statistische Jahrbücher für das Deutsche Reich (Statistical Yearbook for Germany), 1928 bis 1933 (1928 to 1933).

44 Lastkraftwagen in der Volkswirtschaft, Institut für Konjunkturforschung, Berlin, weekly report of 7 May 1930, p. 23–24.

45 K. Mellerowicz: Der Stand des gewerblichen Güterfernverkehrs, in: Die Strasse, issue 5, 1935, p. 145.

to more quickly procure and deliver goods compared to larger trucks. The following table shows the increase of trucks in Germany between 1922 and 1930.[46]

Year	Number of trucks	Growth in %
1930	157,432	9.4
1929	143,952	18.2
1928	121,765	20.6
1927	100,969	12.2
1926	90,029	12.0
1925	80,363	32.5
1924	60,629	17.2
1923	51,736	18.4
1922	43,711	

Table 3: Number of trucks in Germany, 1922 to 1930

This rapid growth in registered motor vehicles led to a greatly higher traffic density and placed an excessive burden on country roads. The national traffic census that was conducted on 10 277 routes in 1928/1929 showed an average number of 319 vehicles per day and, per road for state roads and for provincial roads – the major category for long-distance traffic – 67 horse-drawn vehicles, 64 motorcycles, 142 automobiles and 46 trucks.[47] On these country roads, traffic rose significantly between 1924 and 1928: vehicle traffic, excluding trucks, jumped 210 per cent compared to a traffic census taken in 1924/1925 and truck traffic increased 118 per cent.

During the 1920s, buses developed into competition for the *Reichsbahn* (German State Railroad). While the expansion of the railroad system was brought to a standstill, the network of bus routes expanded to 56 000 kilometres. It broke even with the railroad network and provided the traffic infrastructure for weekend and excursion trips to areas unreachable at the

46 Statisches Jahrbuch für das Deutsche Reich (Statistical Yearbook for Germany) 1924/25, p. 114; 1926, p. 104; 1927, p. 125; 1928, p. 158; 1929, p. 160; 1931, p. 146.

47 Karl Hoffmann: Die Ergebnisse der 2. deutschen Verkehrszählung von 1928–29, in: Verkehrstechnik, issue 47, 1931, pp. 189–191.

time by trains. In order to remain competitive, railroad operators were offering package tours at a reduced fare by the end of the 1920s.[48] Along with the boom of modern means of transportation, tourism as an independent economic factor, was also developing in the 1920s. Many communities became aware of the importance of this economic factor and discovered the active promotion of tourism as a new political field. It was not surprising therefore that the discussion about tourism in the 1920s was linked to discussions on the need for car and bus transportation on the a*utobahns*. The topics 'tourism' and 'a*utobahn*' were referred to over and over again in the newsletter of the Hafraba.

Several times during this period, the inadequacy of the existing country road network was pointed out. Adolf Elsaesser, a road administration official from the city of Mannheim, pointed out in an article that six of the 16 km of the country road between Heidelberg and Mannheim weaved through towns, necessitating lower travel speeds and extra precautionary measures to avoid collisions with railroad cars.[49] In publications concerned with country road and automobile traffic, the conflicts and dangers were repeatedly stressed. For instance, there was the problem of fast motor vehicles having to negotiate the roadways with slower traffic such as pedestrians, cyclists, livestock herds and horse-drawn vehicles. Consequently, speed limits were introduced for automobiles.[50] Carl Kappler proved the specific backwardness of the German country road system after experiencing a 30 000 km test drive across Europe. He was able to show that other European countries had extended the width of their country roads to be two to three meters wider than German roads – and that local authorities had cut down on the practice of cramping trees along the roads, as well.[51] Truck drivers regularly faulted the heavily curved sections on many country roads which caused the vehicles to tilt and shifted the loaded goods. Many a driver would have to restock his goods, thus delaying the delivery.[52] However, in Elsaesser's reports about the condition of country roads, empirically proved,

---

48 Christine Keitz, p. 78s.

49 Städte an den Hafraba-Strassen (Cities along the Hafraba Roads) –III. Mannheim, in: Mitteilungsblatt, issue 5, 1930, p. 7–9.

50 The National Decree on Heavy Vehicle Traffic from 16 March 1928 specified in the first paragraph as road usage in addition to: Riders (Passengers), marching formations, processions, pulled or led animals, in: RGB, I, 1928, p. 66.

51 Carl Kappler: Landstrassen in Deutschland, in: Mitteilungsblatt, issue 9, 1930, pp. 1–2.

52 Julius Meyer and Heinrich Wilkin: Die Landstraße als Heimat des Lastzugbesitzers, in: Die Straße, 1935, pp.160–163.

information on the maximum average speed for specific routes is missing. The Hafraba indicated an average speed of 45km/h in its reports [53] but this speed seems quite high considering the hindrance by unhurried vehicles.

According to Mom, the most important motive for expressway construction was the prospect of greater road safety in comparison to travel on country roads although,[54] remarkably, this argument was barely vocalized in the discussions of the 1920s. The Hafrabra – primarily as justification for building an a*utobahn* system – also commented on the feasibility of driving at a higher speed on *autobahns*.[55] The association's argument was that the a*utobahn* would be the appropriate type of road for motor vehicles – an argument that was later also emphasized by Nazi propaganda. Reports and statistics on road accidents began to appear in the journal '*Verkehrstechnik*' ("Transportation Technology") in 1926 but these never drew any conclusions about the need for an autobahn. The high vehicle density prompted first questions over road safety at major conferences in the United States during the 1920s.[56]

The fragmentation of responsibilities for road planning and road construction into the many different federal states, communities, and provinces contributed to the inadequacy of the road network. In Prussia, for example, the responsibility was transferred from the communities to the Federal States and Provinces on 1 April 1927. A highway code was not uniformly regulated or enforced by the Reich.[57]

---

53 First meeting of the committee on traffic policy at the Hafraba's meeting rooms in Frankfurt/Main, Savignystr. 7, on Monday, 16 May 1927, Schrift Nr. 7 – Hellerdruck Frankfurt 1927, p. 20 (cf. appendix 1).

54 Mom, op. cit.

55 Piero Puricelli, in an essay in an essay in German for the Hafraba, compares the exhilirating speed on the Autobahn with the words „The flight of wild animals", Roads and Car-Only-Roads, in: Mitteilungsblatt, issue 9, 1929, p. 1.

56 See the 2nd Congress on Transportation Security in Washington D.c. with 1,200 participants in March 1926, Verkehrstechnik, 1926, p. 343.

57 Friedrich Frey: Das Hessische Strassenwesen Mainz 1928, Also: Friedrich Frey: Das Neue Strassenverkehrsrecht im Reich und in den Ländern. Mainz 1928. Frey's writing is the only of its kind in the 1920s on traffic laws, which signals the beginning of attention given to the subject. Later, the general inspector for German roads, Fritz Todt, followed suit, in 1932, with his comprehensive study, „Road construction and management," which considered the splintering of road management. See BA, R 4601, File 3001/A.

5 The Work of the STUFA

Controversy and discussions on how the existing road system could be adjusted to serve the rapidly growing traffic of motor vehicles continued. A concept aimed at extending the existing country roads had been put forth. In 1926, the "*Studiengesellschaft für Automobilstraßenbau*" (Stufa – Research Association for the Construction of Automobile Roads) published a blueprint for a road network covering the whole of Germany with a total length of 10 630 km.[58] A route Kassel–Fritzlar–Fulda–Würzburg, which was important for the State of Hesse, was included in this network. The version for this route was planned by the Nazi a*utobahn* project officials in the following order: Kassel–Bad Hersfeld–Fulda–Würzburg. But the project was discontinued at Bad Hersfeld. The network is shown in Image 4.[59] A glance at the rational planning criteria on which the network of the STUFA is based, is worthwhile. The STUFA identified economic areas that produced agricultural and industrial products – as well as raw materials and linked those – as starting and arrival points – along a road network. The planning, thus, was based on the expected economic traffic of goods and individuals. This kind of rational planning was the basis for the traffic policies of the German Reich in 1933 but was abandoned completely during the Nazi era.

A total of 138 roads were identified and numbered accordingly (these are still valid today). Standard procedures for their extension were laid out according to which the standard width would measure six meters. Two-and-a-half meters were allotted for each additional traffic lane. The federal states committed themselves to conducting this extension according to the requirements of traffic development and according to their financial capability, but they stopped short of setting a date for the beginning of work. In 1930, the federal government agreed on a plan – together with the states – for a homogenous extension of the expressway

---

58 Stadtarchiv Kassel (File) 10/3 Hafraba. Also: Verkehrstechnik, issue 18, 1926, K. Heller: Das Kraftwagenstrassennetz Deutschlands , in: Strassenbau und Strassenunterhaltung, supplement to the journal Verkehrstechnik, 1928, p. 669–671. Also: Roland Gabriel, op. cit., p. 30–33.

59 Rappaport, op. cit. The stretch linking Kassel–Fitzlar–Fulda, in this design, occupied only rank 2 in construction (a thin line), but in the STUFA Network of 1928 was increased to rank 1, see Heller op.cit.

network. The total usable width – including the shoulders – was to be increased to eight meters, in mountainous terrain to seven meters.[60]

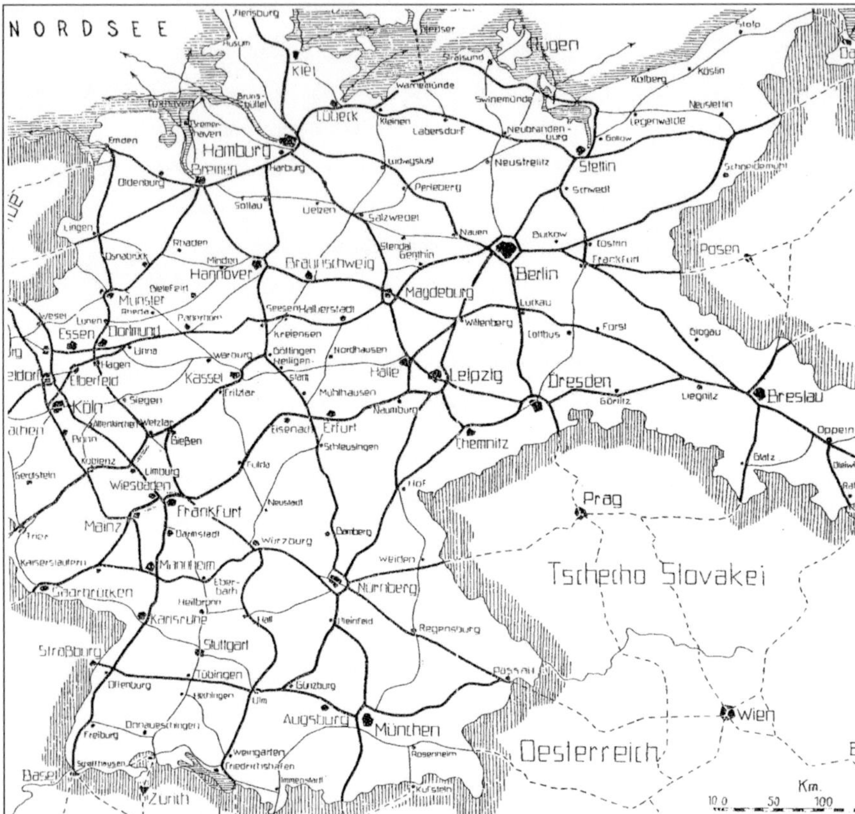

Image 4: The STUFA-Network of long distance roads

60 Ausbau der Fernstrassen in Deutschland, in: ADAC-Motorwelt, issue 38, p. 27–29.

6 The Controversy: *Autobahn* versus Roads

There were plans to construct expressways exclusively for motor vehicles. The promoters of the *autobahn* argued that only this approach could rid the road of slow-moving traffic. Separating the traffic would require expressways that are junction-free and other traffic lines would have to cross them via bridges. Only this design would enable automobiles to travel large distances at high speeds. In a newspaper article entitled "What do you think about motor vehicle-only highways?" a county director by the name of Dr. Usinger, called for the separation of the motor vehicle traffic from all other traffic.[61] Another argument proposed that "drive-through traffic" (cf. Image 5) could be eliminated from the cities with the help of this *autobahn* concept and help preserve buildings through a lessening of ground tremors.[62]

The *autobahn* promoters pursued both local and long-distance traffic concepts. Although Germany did indeed have a very low per capita car density compared to other countries,[63] local traffic was comparatively high and put great stress on country roads. This was the main reason behind the urgent demand for expressways to connect neighboring cities. The idea of an *autobahn* to link the cities of Düsseldorf, Cologne and Bonn was widely being discussed, cf. Section 9. And in 1927, the LEHA (Leipzig–Halle) association was founded with the goal of connecting the cities of Leipzig and Halle by means of an automobiles-only expressway. Discussions on linking Heidelberg with Mannheim via an *autobahn* had already begun in 1926.[64]

---

[61] In: Hafraba-Mitteilungsblatt, issue 3, 1929, p. 1–2.

[62] On the argument of damaging buildings, see Mitteilungsblatt, issue 1, 1928, p. 8; or: Verkehrstechnik, 1929, p. 204.

[63] The density of vehicles per 100 inhabitants in Germany was only one fourth that of France and one third of England's density, see Statistisches Jahrbuch für das Deutsche Reich, 1933, p. 84*. In Italy, the density was only half that of Germany, see H. Uhlfelder: Die oberitalienischen Autobahnen, in: Hafraba-Mitteilungsblatt, issue 7, 1930, pp. 9–14, here p. 11.

[64] Gabriel, op. cit., p. 42–43.

Image 5: Provicial Road Bonn–Cologne, drive-through traffic in Wesseling[65]

The Hafraba association focused on the long-distance Hamburg to Basel (see section 8) *autobahn* route. On 11 January 1927, a conference on expressways was held in Leipzig and its main topic was the a*utobahn* route Berlin–Leipzig–Munich (with a connection to Rome via the Brenner Alpine Pass). The conference focused on modernizing country roads versus building *autobahns*.[66] *Autobahn* proponents however, were isolated at this meeting. Only the representative of the Hafraba, Professor Robert Otzen, stood by the a*utobahn* concept without reservations.[67]

In order to stress the urgency of modernizing country roads, opponents of the a*utobahn* concept insisted that there was no need for long-distance traffic. An a*utobahn* was seen as a

---

65 Die Kraftwagenstrasse Köln–Bonn, edited by Rheinische Provinzialverwaltung, Düsseldorf, 1932, p. 10, Stadtmuseum Bonn.

66 Deutscher Verkehrsdienst, published by Reichszentrale für Deutsche Verkehrswerbung, Nr. 289 from 11 December 1926, Stadtarchiv München (City archive of Munich), Verkehr 84, as well as: Autofernstraße Berlin–Leipzig–München–Rom, Report from the Autofernstraßen-Conference Leipzig, Neues Rathaus, 11 January 1927, Leipzig, Verlag Rats-Verkehrsamt, 1927.

67 Report by Vilbig on the conference in: *Vilbig*: Ausbau der Straßen – Nur-Autostraßen? Drei Musterbeispiele aus Bayern, in: Verkehrstechnik, issue 21, p. 327. Vilbig took part in the conference on the *Autobahn* in Leipzig in 1927.

route serving local or excursion traffic from the major cities with a range of up to 100 km at most. The function of the expressways for easy access to recreation areas near the large cities was widely accepted in the 1920s. It was assumed that an *autobahn* would be less suitable for long-distance traffic in comparison to the soon-to-be introduced air planes and the existing railways. At the time, it was unfathomable that the network of expressways would lead to a great demand for long-distance traffic. At the conference on expressways in Leipzig in 1927, one speaker, Secretary Arthur Speck of the Federal Administration of Saxony, doubted whether there was empirical evidence at all for long-distance traffic on the existing country roads and he applied this argument to the prospective *autobahn* routes.[68] Arthur Speck was an important opponent of the *autobahn* concept who had led the nationwide traffic counts of 1924/1925 and 1928/1929. Secretary Vilbig of the Federal Administration of Bavaria also supported the point of view that modernizing the country roads network be given priority and that the *autobahn* projects should be shelved. Only as an exception, it was reasoned, should an *autobahn* be built to function as an arterial road leading out of cities, or for excursion traffic such as the projected Munich–Starnberg route, cf. Image 62.[69] The *autobahn* opponents viewed expressways as “unrealistic” with a view to the road and railroad networks already in place. An autobahn network, maintained the opponents, was, in fact, beyond human capability and rather illusionary.[70]

---

68 Lecture on ministerial consultant Speck of Dresden, in: Autofernstraße Berlin–Leipzig–München–Rom, Report from the Autofernstraßen-Tagung Leipzig, Neues Rathaus, 11 Januar 1927, Leipzig, Verlag Rats-Verkehrsamt, 1927, p. 16s. See also: Arthur Speck: Der Fernverkehr auf deutschen Landstrassen, in: Verkehrstechnik, issue 13, 1932, pp. 239–242. Speck was opposed to the Autobahn because he found it more urgent to extend the then-existing road network. See also *Gabriel*, op. cit., p. 36s.

69 Vilbig: Ausbau der Straßen – Nur-Autostraßen? Drei Musterbeispiele aus Bayern, in: Verkehrstechnik, issue 21, 1927, pp. 327–330.

70 Rappaport, op. cit., p. 288.

7 The Debate on Financing the Road Network

Financing was a further focus of controversy. The research association for financing German road construction estimated the financial requirements for modernizing the road networks at 3.5 billion RM (*Reichsmarks*). It was proposed to use bonds to raise this money.[71] The net debt services for these bonds were to be covered by fuel and vehicles tax revenue. Opponents of the a*utobahn* concept feared that building these roads would draw funds away from an urgent modernization of the country roads.

In December 1927, the German Parliament debated how road traffic could most suitably be taxed and how the resources could be spent on the extension and maintenance of the road network. Parliament commissioned the government to draw up a memo entitled "*Über die Besteuerung der Kraftfahrzeuge*" (On the taxation of automobiles). This study was released in December 1930 and confirmed the expenses which had been determined by the federal states for road maintenance and new road construction.[72] Table 4 illustrates the commission's findings.

Year	Length of the Network in km	Maintenance in Mio. RM	Basic renewal and improvement in Mio. RM	New Construction in Mio. RM	Sum in Mio. RM	Financed by loans in Mio. RM	Covered by revenue in Mio. RM
	2	3	4	5	6	7	8
1927	269 735	276	261	116	656	210	443
1928	273 390	306	223	118	651	158	490
1929	276 494	326	244	139	713	188	522

Table 4: Expenditures of the German States for Maintenance and New Constructions of Roads

As seen here, the financial requirements are covered at a rate of 25 to 32 percent by foreign-issued bonds. The financial requirements for new constructions were funded by bond

71 Die Finanzierung des deutschen Strassenbaus Berlin 1930. The Association for the Study of Financing German Road Construction was founded in October 1928 by various banks, economic organizations and industrial leaders.

72 Report of the Reichstag, 23 December 1930, 5th election cycle, addendum to the stenographic report, Vol. 449. All following data on taxation are from the memo noted. In column 3 for the current dialogue on road networks all expenses for redemption and interest payments of the laying of road construction are included.

revenues. In subtracting the bond revenues from the overall cost, an overall financial requirement of 433 million RM is determined for 1927 which increases to 522 million RM in 1929. These sums had to be covered by tax revenues. Automobile taxation raised only 156 million RM in 1927. This amount increased to 209 million in 1929. According to §41 of the "*Finanzausgleichsgesetz*" (revenue sharing law), revenue from automobile taxation must be remitted to the federal states for the purpose of maintaining public roads.[73]

The government had strongly increased revenue from the automobile business owing to a rise in customs duties and the purchasing restrictions for grain alcohol in 1930. It announced no further financing measures for the road network. In a law passed on 15 April 1930 relating to the modifications of the customs duty, the tariff for the import of mineral oil was raised from 6 *Reichsmarks* per one hundred kilograms to 10 RM. Previously duty-free gasoline had also been slapped with a duty of 10 RM per one hundred kilograms. For domestically produced mineral oils, a sales equalization tax of 3.80 *Reichsmarks* per one hundred kilograms was approved. In 1931, road traffic faced an increased financial burden in part because domestic fuel manufacturers – as well as importers – were forced by the federal government to mix 3.5 percent of grain alcohol in each gallon gasoline. Overall, a financial burden from duties amounting to 9.55 *Reichspfennig* per litre of gasoline arose. It was estimated that in 1931, tax revenue from automobile taxation amounted to 230 million RM and tax and duties for gasoline amounted to 177 million. In all, 407 million RM would accrue to the federal government from road traffic revenues. However, the advisory board for motor vehicle taxation, which belonged to the interim Reich Council, estimated the financial burden of road traffic at 450 million RM for the fiscal year 1930/31 (in which the limit for the burdening of automobile sales was already exceeded).[74] It follows that the political debate on traffic was strongly influenced by the taxation of fuel and vehicles toward the end of the 1920s.[75] Numerous articles were published in the media, especially in the weekly magazine '*ADAC-Motorwelt*', which complained about the high tax burden on car holders. The government was blamed for thwarting the automobile industry, which in turn would contribute to an economic

---

73 RGB, I, 1927, p. 91

74 Memo, op. cit.

75 Edelmann, op. cit., pp. 140–145, *Reiner Flik*, op. cit., pp. 71–92.

downswing. The journal *Auto und Wirtschaft* reported that the tax burden of the motor traffic industry rose from 275 million RM in 1929 to 466 million RM in 1931.[76]

The argument that building an *autobahn* would deprive the country roads network of government funding for their modernization was countered by the Hafraba, which offered that their plan would not be financed by the government, but rather via private funding in the form of foreign loans and refinancing by toll income. The German Reich was merely to assume a loan guarantee.[77] Herewith, the Hafraba plan met with a legal conflict in connection with the *Finanzausgleichsgesetz*, which did not allow a refinancing of public roads by toll fees. This naturally provided a handy argument for the opponents of the *autobahn* project who were operating from within the Federal Ministry of Transportation.

---

76 Dr. H. Kruse: Die Notlage der Kraftverkehrswirtschaft, in: Auto und Wirtschaft, Nr. 1, Berlin 1932, p. 1.

77 1. Sitzung des Finanzierungs-Ausschusses am Donnerstag, Frankfurt 1927 (cf. Appendix 1).

8 The Hafraba Visions

8.1 The Foundation of the Hafraba in Frankfurt (Main)

Among the other initiatives to expand the road network, the Hafraba, or "Association for the Preparation of the *Autobahn* of Hanseatic cities–Frankfurt–Basel", the counterpart to the STUFA, was the most important one.[78] It was founded in the town hall of Frankfurt (Main) on 6 November 1926 and played a significant role in Hessia's regional history.[79] The abbreviation HAFRABA signifies Hamburg–Frankfurt–Basel (later on: Hanseatic cities–Frankfurt–Basel) and denotes the aim of the association: to promote an a*utobahn* from the ports on the North Sea (i.e. Hamburg), via Frankfurt (as an important commercial hub in western Germany) on to southern Germany at the border of Switzerland, i.e. Basel.

Image 6: Cover of the Statues of HAFRABA 1929

[78] "Frankfurt" refers to "Frankfurt am Main" in the following.

[79] Report of the first meeting of the board of Hafraba in Frankfurt on 10th February 1927 (cf. appendix 1).

The Hafraba foundation was preceded by a meeting in Frankfurt on 9 September 19 26 for which the local department of economics in Frankfurt had dispatched about 70 invitations to interested cities, provinces, associations, and business corporations.[80] Frankfurt portrayed itself as a 'city of streets'. During the introductions, Councilman Schlotter remarked that Frankfurt constitutes an important junction for German and European roads. He suggested bonds at the international capital market in order to finance the project. Additionally, Schlotter emphasized the argument, that was taken on again and again by the Hafabra in later years, that the construction of high-capacity roads for automobiles would attract foreign tourists.[81] He envisioned that American tourists would arrive with their cars at the shipping ports of Hamburg and Bremen and reasoned they should be provided with a high-capacity road to the Mediterranean Sea. Schlotter went on to alert the meeting's participants of French plans for a road from Cherbourg to Nice that would shorten the way to the popular Mediterranean destination by about 400 kilometres compared to the way via Hamburg. This would obviously have a detrimental effect on tourism in Germany. Comparing those arguments with data on tourism, it can be seen that the number of American tourists increased from 76 000 in 1926 to 159 000 in 1932. However, the proportion of foreign tourists who visited Germany remained almost steady – between 13 and 15 per cent. The number of automobiles carrying American tourists, i.e. that were unloaded at the ports, could not have been high because only 474 vehicles from the USA were reported in Germany in 1926/1927.[82] The argument that an a*utobahn* network would attract foreign tourists and generate foreign currency revenue was emphasized by Fritz Todt, the leader of the Nazi a*utobahn* project, in later years.[83] To justify the planning of an expressway from Turin to Savona, the argument

---

[80] Bericht über die Besprechung des Projekts Automobilstrasse Hamburg–Mailand, Druckschrift des Wirtschaftsamts Frankfurt, 1926, Stadtarchiv Kassel, Akte Städtisches Verkehrsamt, 10/3 Kassel City Archive, City Traffic Office file, 10/3

[81] Wilhelm Cuno: Die Hafraba und der Fremdenverkehr von Übersee, in: Mitteilungsblatt, Heft 4, 1929, p. 2s, also: Thalenhorst: Städte an den Hafraba-Strassen: Bremen, in: Mitteilungsblatt, Heft 4, 1930, pp.1–3.The Hafraba distributed a brochure, which included the article from Marcel Violett: What the French want, from L'Echo des Sprots (undated, likely from 1927), in German, containing the Cherbourg–Nice project as the French response to the Hafraba design, city archive of Kassel, traffic office file 10/3. The contest between Germany and France to attract American tourists gained more and more attention within the Hafraba, see for example Frey's, "Strassenwesen und Strassenverkehrswesen" in „Rhein-Main-Gebiet", in: Mitteilungsblatt, Heft 11, 1929, p. 6.

[82] Statistisches Jahrbuch für das Deutsche Reich, Jg. 1927, p. 48, Jg. 1933, p. 51, Jg. 1928, p. 159.

[83] Fritz Todt: Reichsautobahnen in der Wirtschaft, Frankfurter Volksblatt, 27 February 1936, Archiv Hessischer Rundfunk (Hessian Radio), Zeitungsausschnitte Autobahnen.

that foreign tourists would be attracted was again stressed in the 1950s and 1960s by inferring that tourists from northern Europe would need this connection to the Mediterranean Sea.[84]

From the beginning, the association planned to extend the a*utobahn* from Basel to the Italian port of Genoa, targeting automobile drivers heading to the Mediterranean Sea and Nice. Image 7 shows the projected course of the *autobahn* with the turn-off to Bremen.[85] In Switzerland, the "Association for the Promotion of the Construction of the *Autobahn* Basel–Italian border" was founded to promote the extension from Basel to the Italian border.[86]

Image 7: The North-South Route of the Hafraba Autobahn through Germany with its turn-off to Bremen

84 Michele Bonino and Massimo Moraglio: Inventing Movement – History and Images of the A6 Motorway, Turin 2006, p. 79s.

85 Robert Otzen: Die Autostrasse Hansestädte–Frankfurt–Basel, Ziele und Zweck des Vereins "Hafraba", Hannover 1927, p. 7 (cf. Appendix 1).

86 Mitteilungsblatt, issue 7, 1929, p. 11; report on the exhibition of the Hafraba in Basel, in: Baseler Nachrichten on 31 August 1927, City Archive Kassel, Akte Städtisches Verkehrsamt, 10/3.

When one explores the political and social forces that were brought together in the Hafraba association, one has to conclude that there were influences emanating from the construction and transport industries. These were complemented by the transport administrations of the states and cities. Therefore, the Hafraba association can be seen as a mix of entrepreneurs and public agencies ("public-private-partnership"). It repeatedly claimed to be a private initiative – and not part of a state administration and set itself apart from the LEHA association which it denounced as "bureaucratic".[87]

The strong participation of municipal institutions in the Hafraba caused concern that a lot of public money would be used for a*utobahn* projects. The Upper-Hessian Automobile Club of Giessen, for example, was uneasy about the prospect and stated – on 5 June 1928 to the city's Chamber of Commerce and Industry – that *autobahn* projects should be prevented from receiving public money. The case was made that it would be impossible to raise sufficient private capital without public help. Public money, this club reasoned should be used for more practical purposes than for building an a*utobahn*.[88]

There were two groups among the Hafraba association's 32-member advisory board and six-member executive board.[89] One group of members consisted of:

- leading officials – including mayors – of the urban administrations of the cities along the route from Hamburg to Basel: Bremen, Hamburg, Hannover, Göttingen, Kassel, Giessen, Marburg, Mainz, Frankfurt (Main), Wiesbaden, Darmstadt, Mannheim, Karlsruhe and Freiburg. The mayor of Frankfurt, Dr. Landmann, played a leading role in the Hafraba, providing the city hall conference rooms several times for meetings.
- leading officials of the transport administrations of the states along the Hamburg–Basel route, e.g. Arthur Zierau, representing the Federal State of Baden, Kurt Becker from the Prussian province of Hesse-Nassau and Dr. Krebs from the State of Hesse-Volksstaat.
- Robert Otzen, Professor of Railroad Construction from the Technical University of Hannover. Otzen was chairman of the Hafraba until 1931.

---

87 Report of the first meeting of the advisory board, op. cit., p. 10.

88 Hessisches Wirtschaftsarchiv Darmstadt, 9/1749.

89 The list of the board members is published in Report of the first meeting of the advisory board of Hafraba on 10 February 1927, op. cit., p. 28–29.

The second group consisted of ten directors from the construction and transport industries, for example, Vogel and Dykerhoff from the *Portland Zementwerke* (cement factory) in Amöneburg and Heidelberg, Milke from the *Strassenbau AG* in Cologne and Meisenheimer from Wygass+Freytag in Frankfurt (Main). Five other representatives came from chambers of commerce. The spokesman for the advisory board was Landmann, the mayor of Frankfurt (Main). The secretary of the Hafraba association was Willy Hof. At the end of 1929, the Hafraba consisted of 125 members, representing 12, 13 (12 of 13?) provinces, 33 towns, 50 companies and six chambers of commerce.[90]

One of the most prominent members of the Hafraba executive board was the Italian highway promoter Piero Puricelli. He spoke German fluently and served as an important link between Italy and Germany. This exchange has been explored in great detail by Bortolotti.[91]

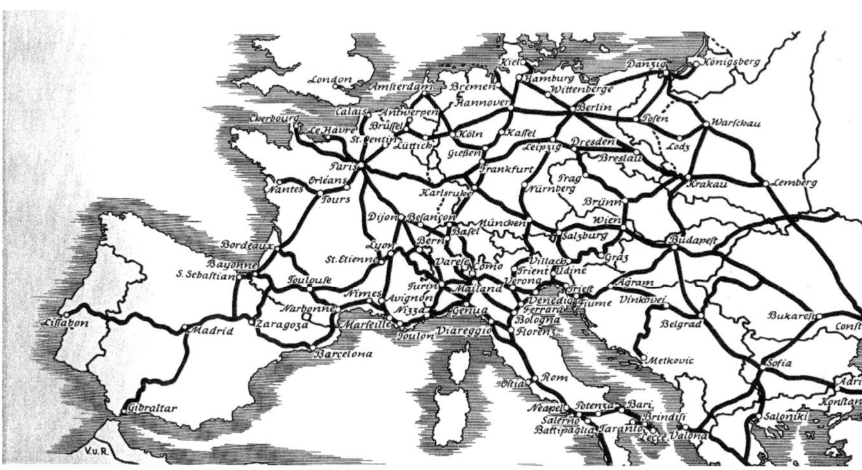

Image 8: Puricelli's draft of a European Autobahn network[92]

90 Robert Otzen: Drei Jahre Hafraba, in: Mitteilungsblatt, issue 11, 1929, p. 3.

91 L. Bortolotti: Fra Politica, Propaganda e Affari: L'Autostrada Roma–Berlino 1927–1942, in: Storia Urbana, Vol. 81, 1997, pp.47–79.

92 Puricelli, P.: Entwurf für ein europäisches Autostrassennetz, in: Die Strasse, Heft 2, 1934, p. 42s. Puricelli presented his ideas at the 7th international congress for road construction in Munich from 3–8 September 1934.

The speaker of Hafraba's executive board, Robert Otzen, often underscored the strong impetus of Puricelli which led to the foundation of the Hafraba association.[93] In 1926, Otzen reported on an excursion which he made as Puricelli's guest on the Italian expressway and of his first impression of a motor vehicle ride on a road without obstacles.[94] At the first meeting of the advisory board in Frankfurt in 1927, Puricelli gave a speech in German, emphasizing[95] not only the idea of creating a modern road network in the form of the German a*utobahn*, but also the concept of utilizing both private enterprise and public agencies in order to bring the idea to fruition.

93 Robert Otzen: Die Autostrasse Hansestädte–Frankfurt–Basel, Ziele und Zweck des Vereins „Hafraba“, Hannover 1927, p. 15.

94 Report of the first meeting of the board of Hafraba on 10 February 1927, op. cit., p. 13.

95 Op. cit.

HAFRABA

Mitteilungsblatt des Hafraba e.V.

/EREIN ZUR VORBEREITUNG DER AUTOSTRASSE HANSESTÄDTE-FRANKFURT-BASEL

›as Mitteilungsblatt wird an ›itglieder, Verwaltungsrat ›nd Ausschüsse kostenlos ›rsandt. Abonnementspreis: 6 RM. im Jahr Einzelnummer 50 Pfg.	HERAUSGEGEBEN VON DER GESCHÄFTSSTELLE FRANKFURT A. M. Tel. Adr.: Hafrabastraße / Leerbachstr. 112 / Tel.: Zeppelin 55367	Erscheint monatlich Bezug durch die Post Nachdruck nur mit Quellenangabe gestattet

›ummer 9 — 1. September 1931 — 4. Jahrgang

Aufnahme einer Sitzung des I. Internationalen Autobahn - Kongresses im Verwaltungsrats-Sitzungssaal des „Internationalen Arbeitsamts in Genf"

I. Internationaler Autobahn-Kongreß in Genf

(Sonderbericht für das »Hafraba-Mitteilungsblatt«)

Der Kongreß tagte vom 31. August bis 2. September Gebäude des „Internationalen Arbeitsamtes" in Genf. ›ffnet durch eine Ansprache des Direktors des Internationalen Arbeitsamtes, Herrn Albert Thomas, der : Gründung eines Autobahnkongresses im Interesse › Behebung der internationalen Krise sehr begrüßte, ›n dieser sei nicht nur in verkehrstechnischer, sondern › allem auch in wirtschaftlich-sozialer Hinsicht von Be›tung. Die Tatsache, daß der Kongreß im Verwal›gsratssaal des Arbeitsamtes tage, sei schon Be›s dafür, welch persönliches Interesse Herr Thomas ›bst an den aufgeworfenen Fragen habe und wie eng die Beziehungen, die zwischen dem Zweck der Tagung und den Zielen seines Arbeitsamtes beständen, seien. Er ermutigte die Teilnehmer an dem Kongreß, vor der öffentlichen Meinung die Begründung ihrer Bestrebungen darzulegen und ihre Ideen zu entwickeln, wobei in den gegenwärtigen Zeiten auch der moralische Faktor betont werden dürfe, denn hier seien ja Mittel zur Bekämpfung der Arbeitslosigkeit gegeben, um so mehr, da der Gedanke der Autobahnen geeignet wäre, für das internationale Kapital eine Verlockung zu sein.

Zum Ehrenpräsidenten des Kongresses wurde Herr Dr. Ing. Puricelli, Königl. Italien. Senator, Er-

1

Image 9: Layout of the Hafraba newsletter

8.2 The Activities of the Hafraba

The Hafraba association published numerous papers to promote its ideas. By 1927, eleven publications had appeared (see appendix 1.) From 1928 on, the newsletter "HAFRABA – *Mitteilungsblatt des Hafraba e.V.*" was published in twelve issues per year. Image 9 depicts a front page of the newsletter whose layout was in the functional style of the 1920s. Starting in 1932, the newsletter was entitled "*Die Autobahn"*. In its newsletters, the Hafraba also published information from organizations that it cooperated with, such as the LEHA association and the Swiss *Autostrassen Assoziation* (SVA). In 1933, the editors of the newsletter switched their political alliances to Hitler as he was the most important promoter of the a*utobahn* idea. The July and September 1933 editions showed Hitler on the front pages and the July 1933 edition touted the headline: "The a*utobahn* network is the path to the rebirth of Germany's rise and glory". In 1934, the number of issues nearly doubled from twelve to 22. And when, on 18 August 1933, the Hafraba association was compulsorily dissolved and incorporated into the project company Gezuvor – *Gesellschaft zur Vorbereitung des Reichs-Autobahnbaus* (Company for the Preparation of Autobahn Construction)[96] the Gezuvor went on to publish "*Die Autobahn*". The chief editor of the Hafraba newsletter, Kurt Kaftan, kept his position until the end of 1934,[97] when the journal was discontinued. The head of the Nazi a*utobahn* project, Fritz Todt, published his own journal "*Die Strasse*" – initially in 24 issues per year but with a reduced number of issues from 1940 until 1942. This journal was already in print by August 1934, so that for the remaining months of 1934, both journals – "*Die Strasse*" and "*Die Autobahn*" – appeared simultaneously. From 1935 on, the front covers of "*Die Strasse*" were subtitled "*Die Autobahn.*" At the first glance, there was no difference between the Hafraba and the Nazi visions of the *autobahn.* In fact, "*Die Strasse*" broadened its scope: the NS *autobahn* project expanded the single Hafraba route from Hamburg to Basel to an entire network and regarded the *autobahn* as a monument of the NS policy that would last for hundreds of years.[98] In 1934, a former member of the Hafraba execution board, Walter Becker, provided some unintentional comedy when he compared the construction of the *autobahn* with the pyramids in Egypt.[99] Further, "*Die Strasse*" contained a curious mixture of obtrusive NS propaganda to reports on Hitler's expansionary foreign policy and "war successes" after its degeneration beginning in 1938.[100]

---

96 Die Autobahn, issue 9, 1933, pp. 7–9.

97 In Todt's journal "Die Strasse" Kaftan had no influence. Only once, at the opening of the first *Autobahn* route from Frankfurt (Main) to Darmstadt in May 1935, was he allowed to publish a retrospect: „Hafraba" – Ein Rückblick, in: Die Strasse, issue 9, 1935, p. 321s. This was the only time the HAFRABA project was mentioned in Todt's journal. Kaftan published his memoirs about the *Autobahn* project in 1955: *Kaftan, K.*: Der Kampf um die Autobahnen, Berlin 1955.

98 Statement of the member of the government, Rudolf Hess, in the opening speech of the road construction congress in Munic on 6 September 1934, in: Die Strasse, issue 2, 1934, p. 38.

99 see Becker: "Von der HAFRABA zur ReichsAutobahn" in: Kasseler Post on 18 March 1934, p. 3, UBK.

100 See tables of content of Die Autobahn 1932 to 1934, Hafraba-Mitteilungsblatt 1928 to 1931 and Die Strasse 1934 to 1942 on the web site www.vahrenkamp.org.

In its publications, the Hafraba association unveiled the weaknesses of the existing road network in Germany. It promoted expressways for automobiles by calling attention to the growing demand for more road capacity due to increasing motorization levels and the goal of connecting important cities. The Hafraba association presented a plan for an a*utobahn* network for the whole of Germany as early as 1927,[101] and was later imitated by many other promotional groups. But Hafraba's executive board decided to focus on the original north to south (Hamburg–Frankfurt–Basel) route and not to extend this route into a whole network. When in 1930, the town planning counselor of Wetzlar, Theodor Golder, published a plan for extending the Hafraba route into a road network throughout Germany, the executive board wrote a letter to the mayors of the cities along the Hafraba route and claimed that Golder did not represent the position of the Hafraba.[102] Image 10 shows this map. Fritz Stück, a town planner from Kassel, also proposed several additional routes such as a Cologne–Kassel–Leipzig route.[103] The spokesman of the Hafraba, Robert Otzen, was also a proponent of a complete a*utobahn* network. He asserted that only a network would attract the broad political support required to implement the project.[104] However, the majority of the executive board members did not agree with him, prompting his resignation as spokesman in May 1931. His opponents brought forward the argument that a network would weaken the position of the Hafraba cities in south-western Germany.

On the first meeting of the technical committee on 5 April 1927, Otzen called on the Hafraba to quickly finish the draft for the north-south route (approximately 900 km) so that it could on hand for the first "International Automobile Exhibition" in Cologne which would take place 20–31 May 1927.[105] The draft of the route initially consisted of 45 segments, and later on of

---

101 R. Otzen: Die Autostrasse Hansestädte–Frankfurt–Basel. Zweck und Ziele des Vereins "Hafraba", Schrift Nr. 1, Hannover 1927, p. 13f (cf. Appendix 1).

102 Golder: Weiterausbau der Hafraba-Idee, in: Hafraba-Mitteilungsblatt, issue 2, 1930, pp. 1–3, Stadtarchiv Karlsruhe, Akte 1/H/3500. The Golder plan might have been a prototype for the Nazi Autobahn project.

103 Fritz Stück: Um Kassels Zukunft – das Aufbauprogramm, Kassel, Geotechnischer Verlag 1930.

104 Robert Otzen: Zur Systematik des deutschen Landstrassenbaus, in: Bautechnik, issue 26, 1931, Stadtarchiv Karlsruhe, Akte 1/H/350.

105 Report of the first meeting of the technical committee of the Hafraba association in Frankfurt/Main on Tuesday, 5 April 1927, Schrift Nr. 5 – Hellerdruck Frankfurt 1927, p. 5s (cf. appendix 1).

53, with each stretching along 20 km of the route pictured on a scale of 1:100,000.[106] The following segments were included:

1. Construction section I, Hanseatic cities–Hanover:
   Lübeck, Travemünde, Warnemünde, Lesum, Eichede, Ahresburg, Hamburg, Stade, Carlsdorf, Soltau, Bergen, Winsen, Hanover, Sarstedt, Gronau, Alfeld, Einbeck, Nörten, Bremen, Verden, Walsrode, Hagen.
2. Construction section II, Kassel:
   Göttingen, Reinhardswald, Kassel-North, Kassel-South, Fritzlar, Jesberg, Kirchhein, Marburg counterdraft Marburg
3. Construction section III, Frankfurt–Darmstadt:
   Gießen, Bad Nauheim, Friedberg, Frankfurt, Möhrfelden, Darmstadt, Bensheim.
4. Construction section IV, Karlsruhe:
   Mannheim, Heidelberg, Wiesloch–Speyer, Bruchtal, Karlsruhe, Raststatt, Baden-Baden.
5. Construction section west-east route:
   Frankfurt, Mainz.

Kurt Becker, state chief construction counsellor from Kassel, was responsible for the development of construction section II. This segment consisted partly of four-lane roads; three lanes were set aside for lesser traffic, apparently following the Italian version. Additionally, there was a shoulder of one meter width for each direction. Each lane was planned for a width of three meters. The question of how the two directions would be separated from each other remained unanswered in the draft. Apparently, no middle crash barrier or road marking was planned. For the route Hannover–Kassel–Giessen, presumably low in traffic, three lanes were planned, however, only two lanes would be built in the beginning.[107]

---

106 Robert Otzen: Die Autostraße Hansestädte–Frankfurt–Basel, Erläuterungsbericht zum Entwurf der Autostraße (see above, Explanation for the Design of Motorways), Schrift Nr. 4 – Hellerdruck Frankfurt 1927, p. 4 (cf. Appendix 1). Also: Gabriel, op. cit., p. 45. Hafraba-Mitteilungsblatt issue 1, 1928. Kasseler Post on 29 January 1928, UBK. None of the volumes of the pre-sketch could be found in the archives.

107 Otzen, op. cit., pp. 8, 15 and First meeting of the technical committee of the Hafraba association in Frankfurt/Main on Tuesday, 5th April 1927, Schrift Nr. 5 – Hellerdruck Frankfurt 1927, p. 10.

The Hafraba demanded that the Frankfurt–Mannheim section function as a showcase and demonstration project. In doing so, the Hafraba association already addressed the issue of local transportation with the "small HAFRABA" for the Frankfurt–Mannheim region. These a*utobahn* plans responded to the traffic situation in the Rhine-Main region, which was "unbearable" according to the Darmstadt Chamber of Commerce – allegedly leading to a bottleneck for economic development. The chamber of commerce referred to this situation as a "traffic predicament" insofar as the *Reichsbahn* (national railway system) system did "not even remotely" meet the existing demand.[108] Image 11 shows a map of the planned routes from the year 1928.[109]

In 1931, the Hafraba altered its 1927 concepts and instituted new guidelines. The width of the a*utobahn*, it maintained, must respond to the increasing width of the automobile and stipulated an adjusted width of 20.5 meters.[110] For each direction two lanes were planned of 3.50 m width and in the middle a separation strip of three meters width. This marked a first in a*utobahn* history whereby direction lanes would be separated by a middle strip. This would serve various functions – not only traffic safety. Piwecki commented on this idea: „It's important that the width of the middle strip also provide for the planned possibility of delivering energy and electricity, traffic signs, and, with regard to the free market initiative of the association, possibly commercial billboard advertising."[111] In a conversation on 23 December 1931 in the Hessian Finance Ministry, the Hafraba decided that the planned stretch from Frankfurt to Mannheim would correspond to the new guidelines.[112]

---

[108] Hafraba-Mitteilungsblatt, issue 12, 1929, p. 7.

[109] Institut für Stadtgeschichte Frankfurt/Main, Magistratsakten MA6289.

[110] Uhlfelder: Richtlinien für den technische Anlage von Autobahnen, Teil A: Strecken im Flachland, in: Mitteilungsblatt 4/1931, pp. 4–6, derselbe: Richtlinien für die technische Anlage von Autobahnen, Teil B: Stationen, in: Mitteilungsblatt, Heft 10, 1931, pp. 6–8, derselbe: Richtlinien für die technische Anlage von Autobahnen, Teil C: Abzweigungen und Kreuzungen, in: Mitteilungsblatt, Heft 11, 1931, pp. 2–7.

[111] Frank Piwecki: Frühe Konzepte des individuellen Massenverkehrs in Deutschland, dargestellt am Beispiel des Vereins "HAFRABA", Magisterarbeit im Hauptfach Geschichte, Universität Hannover 1996,.p. 100.

[112] Hafraba-Mitteilungsblatt 1/1931, p. 5.

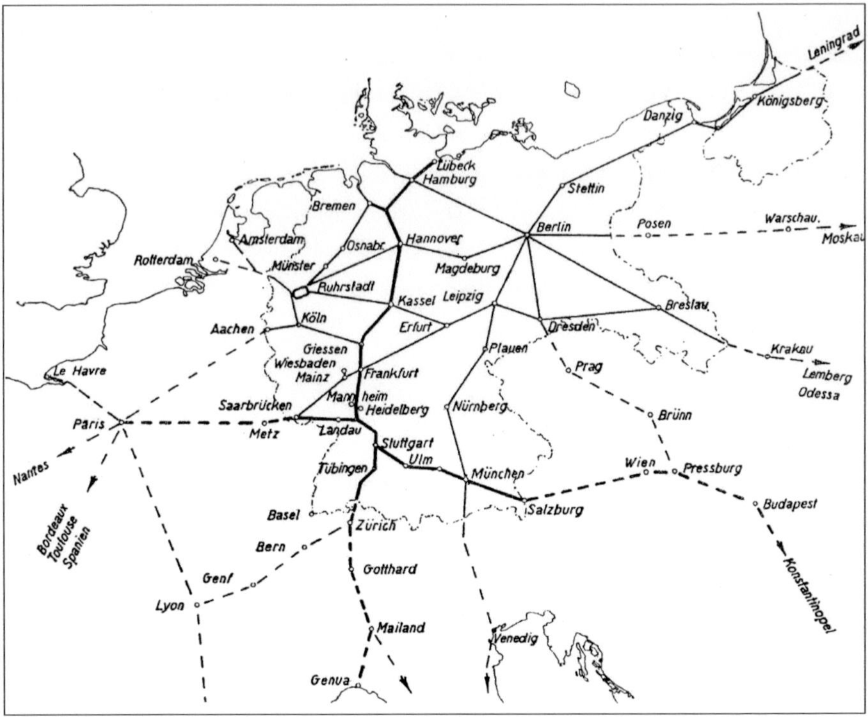

Image 10: Golder's draft of an *Autobahn* network for Germany in 1930

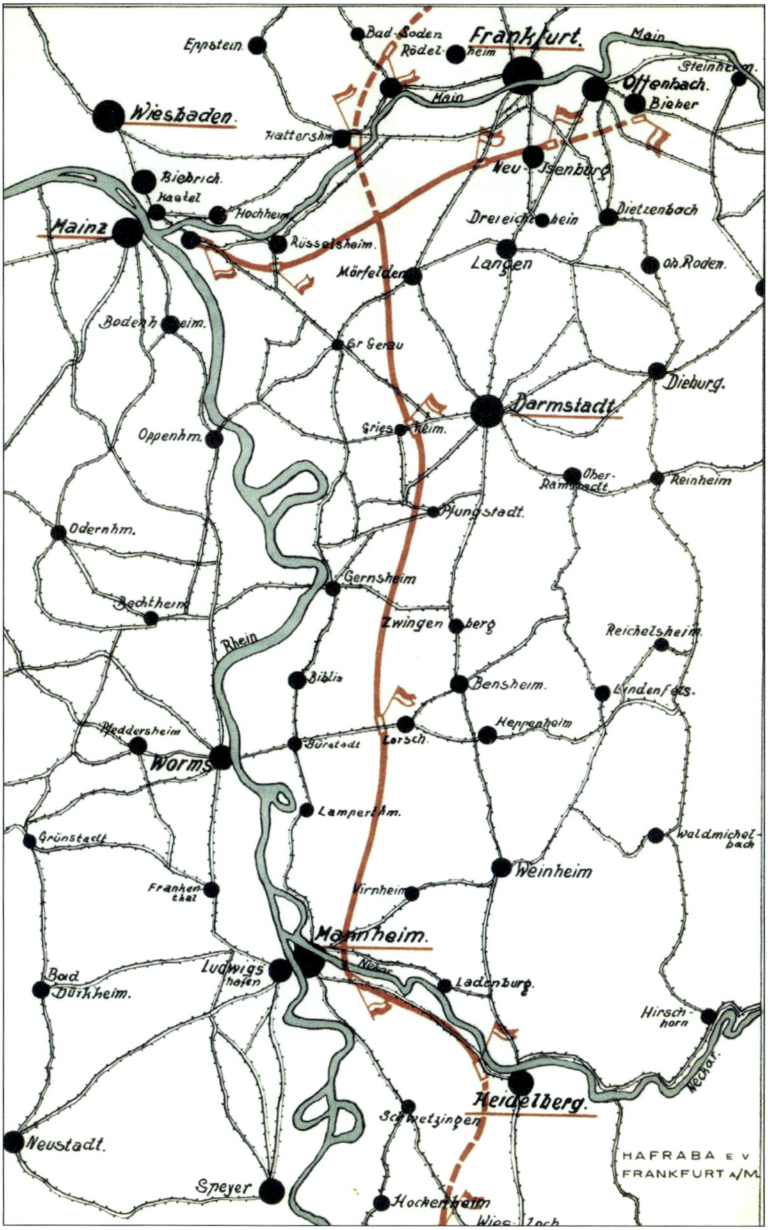

Image 11: The "small" Hafraba Frankfurt–Mannheim route

Besides proposing routes and networks for the a*utobahn*, the Hafraba association also focused on the following subjects in its publications:

1) The question of how to finance the a*utobahn* projects. Italy served as a role-model. The north-south Hamburg–Hanover–Kassel–Frankfurt–Basel route was to be financed by private capital and refinanced by revenue from toll fees. This concept was introduced to the public in a memorandum dated 17 March 1928.[113] However, building privately financed roads was difficult in Germany. This process would be at odds with with §13 of the *Finanzausgleichs-gesetz*.[114] Furthermore, the transport administration did not support privately financed roads, favoring instead the interests of the *Reichsbahn*.[115] The Hafraba applied for an exemption from paragraph thirteen in Parliament and,[116] also petitioned the transport ministry on 28 August 1932, to approve a privately financed Frankfurt–Mannheim–Heidelberg route. Since Arthur Speck, an official of Saxony's road administration, was regarded as a significant opponent of the a*utobahn* construction, the application aimed to respond explicitly to his arguments and to refute them. Four volumes with the planning details were attached to the motion, but none of these volumes have ever been found in the archives.

From the memoirs of Kaftan, Brüning and Treviranus, one gets the impression that the Brüning cabinet in 1932 wanted to immediately put into effect the advanced plans of the Hafraba to construct the Frankfurt–Mannheim a*utobahn* as a job creation measure under the Fifth Urgent Order.[117] This viewpoint, however, is not supported by the minister's discussions in preparation for the Fifth Urgent Order.[118] The talks dealt merely with the construction of

---

113 Stadtarchiv Kassel, Akte Städtisches Verkehrsamt, Hafraba, 10/4.

114 On the problem of financing, see Kaftan, op. cit., pp. 111–124.

115 Record of the conference of the HAFRABA with the Transport Administration on 18 August 1928, in: Thüringisches Hauptstaatsarchiv Weimar, C233, p. 22. See also Hafraba-Mitteilungsblatt, issue 1, 1928, p. 6.

116 Application by representative Rippel et al. from 4th of March 1931, Institut für Stadtgeschichte Frankfurt/Main, Magistratsakten MA6289. Application by Koch-Weser et al. from 4 July 1930, Kaftan, op. cit., p. 124s.

117 Kaftan, in the place cited, p. 133, See also G. Treviranus: Die HAFRABA im Arbeitsbeschaffungsprogramm1932 (Hafraba and Job Creation), in: Bundesminister für Verkehr (ed.): HAFRABA – Bundesautobahnen Hansestädte–Frankfurt–Basel, Bau Verlag, Wiesbaden, 1962, pp. 42–43. Treviranus was member in the Cabinet Brüning. See also: Heinrich Brüning: Memoiren 1918–1934, München 1972, vol. 2, p. 605.

118 For the meeting of the cabinett on 19 May 1932 and 20 May 1932 see document 757 and 759, in: Akten der Reichskanzlei Weimarer Republik, edited by Karl Dietrich Erdmann, Kabinette Brüning I u. II, volume 3, Oktober 1931 till Juni 1932, Boppard am Rhein, 1990, Sp 2539s and 2544s.

streets in general, not with construction of the *autobahn*. There were also influential cabinet members who had spoken openly against the *autobahn* plans. The Federal Finance Minister, Mr. Papen, in a memo dated 30 September 1932 to the Federal Transportation Minister and the Federal Economics Minister, outlined the opposing position of his ministry to the *autobahn* and to its financing with toll fees. With this, he expressly pointed out that the Federal Transportation Minister had refused the parliamentary motion in support of the Hafraba plan during a cabinet budgetary advising session 19 February 1931.[119]

2) The importance of the automobile industry for the economic development of a country. This referred especially to the high number of employees in the automobile industry and its subcontracting industries. Dr. Krebs, an official in Darmstadt's Ministry of Domestic Affairs, for example, reported on a study trip to the USA where he had observed the effect of the car industry on employment rates.[120] Furthermore, traffic on an efficient road network was considered a source of economic growth and an instrument for conducting commerce.[121] Road building was repeatedly referred to as a means of job creation at international road building conferences. For the Frankfurt–Mannheim route, an estimated 6000 people were employed for a duration of two years.[122] The journals *Verkehrstechnik* and *ADAC-Motorwelt* also reported frequently on this topic. According to the plans of the Hafraba, the *autobahn* project would be refinanced by toll revenues. The profitability of the project, thus, played a decisive role. The argument for hiring the unemployed to construct the *autobahn* gained greater importance when it was reasoned that the lower cost of hiring the unemployed would reduce construction costs. This idea was presented at the 2nd International Highway Convention in Milan and also appeared in the chapter "The industrial-engineering importance for the working services" of the Hafraba petition to be granted a certificate of exemption with regard

---

119 A R2/23460.

120 Hafraba-Mitteilungsblatt, issue 2, 1930, pp. 2–6; and Neumann: Strassenverkehr und Strassenwesen in den USA, in: Hafraba-Mitteilungsblatt, issue 1, 9131, p. 4s. On the „model character" of America for the German car industry, see also Heidrun Edelmann: Vom Luxusgut zum Gebrauchsgegenstand, Frankfurt 1989, pp. 69–79. Numerous reports on the USA can also be found in the journals Verkehrstechnik and ADAC-Motorwelt.

121 J. Horn: Autobahn und Industrie, in: Hafraba-Mitteilungsblatt, issue 7, 1931, p. 1s. In Hafraba-Mitteilungsblatt 8/1930, p. 19, there is a report: Das Auto als Konjunkturhebel.

122 Mitteilungsblatt 1/1931. See also „Wege zur Belebung der Bauwirtschaft unter Berücksichtigung der Hafraba", lecture by U. Rouselle, in: Hafraba-Mitteilungsblatt issue 7, 1931, p. 5s.

to the funding of the Frankfurt–Mannheim *autobahn*. The motion was submitted to the Federal Ministry of Transport on 28 August 1932, as mentioned previously here.

3) The Hafraba used modern empirical methods in determining how much traffic could be anticipated for the Frankfurt–Mannheim route. It distributed a questionnaire among German entrepreneurs in cooperation with the Frankfurt Chamber of Commerce and Industry on 11 September 1928.[123] Frankfurt's Chamber of Commerce and Industry at Main-Hanau included a memo with the questionnaire to its members, reminding them of the importance of economical traffic within the Rhein-Main region. This approach is part of the discussions on profitability and rationalization in the 1920s. Parliament established the Enquete Committee in order to highlight, using empirical methods, the situation in individual branches of the economy. The explanation in the questionnaire indicated that an expressway would allow for time saving because the goods could be delivered much faster. In order to demonstrate the advantages of their approach, the Hafraba used the Frankfurt–Heidelberg route as an example in the questionnaire. The distance covered on country roads was 83.5 kilometres compared to just 71 kilometres on the planned road by the Hafraba. Fifty-three per cent of the country road ran through small towns and the roads were described as partly narrow and curvaceous. The speed limit in small towns was 25 kilometres per hour. At an average speed of 31.5 kilometres per hour, the whole route could be safely driven in two-hours and 39 minutes. In comparison, the average travel speed on the Hafraba road was 75 kilometres per hour allowing this road to be mastered in 55 minutes. Consequently, almost 1¾ hours could be saved. A four-colored map was attached to the questionnaire, showing the Frankfurt–Heidelberg route (see Image 11). At the same time, the cross range from Neuisenburg to Wiesbaden was shown, indicating the subsequent a*utobahn* junction in Frankfurt. The main route had exits in Griesheim and Lorch before it reached Mannheim. Despite all of this, however, the success of the Hafraba questionnaire was disappointing. Only 82 of the 2900 dispatched questionnaires were returned. The Hafraba published the traffic census results.[124]

---

123 Institut für Stadtgeschichte Frankfurt/Main, Magistratsakten MA6289.

124 Route Frankfurt/Main–Offenbach–Darmstadt, in: Hafraba-Mitteilungsblatt 2/1929, p. 6. Route Frankfurt/Main–Mainz, in: Hafraba-Mitteilungsblatt, issue 2, 1928, p.6 and issue 8, 1931, p. 9s.

4) Technical questions on the construction of an *autobahn* and the layout of junctions.[125] The junction options "baroque", "renaissance" (clover) and "gothic" were distinct and rather peculiar terms for a modern traffic concept. The clover junction form was created and patented by Willy Sarbach in Switzerland, who had been inspired by an exhibition of the Hafraba in Basel in 1927.[126]

5) Kurt Becker, a member of the Hafraba executive board, in several of the association's magazines championed a specific technique on how the road's edges should be properly built. This topic was taken up by Todt later on and the landscape lawyer Alwin Seifert explored it in-depth by order of Todt. During a speech before gardening executives, Becker showed a poorly constructed bridge on an Italian expressway. The Hafraba disapproved of the purely functional design of bridges and underpasses of Italian expressways. Instead, it favored architectural principles that blend constructions into the landscape. Thereby, it anticipated Todt's concepts of landscape aesthetics.[127] Becker demonstrated his ideas in a lecture for horticulture officials and showed a bridge crossing an Italian expressway as an example of bad style, cf. the following Image.[128]

Image 12: A bridge crossing an Italian expressway

125 First meeting of the technical committee of the Hafraba association in Frankfurt/Main on Tuesday, 5 April 1927, Schrift Nr. 5 – Hellerdruck Frankfurt 1927, also H. Uhlfelder: Richtlinien für die technische Anlage von Autobahnen (Guidelines for the Construction of Motorways), Teil C: Abzweigungen und Kreuzungen, in: Hafraba-Mitteilungsblatt, issue 11, 1931, pp. 2–7.

126 Uhlfelder, op. cit., and K. Kaftan: Der Kampf um die Autobahnen, Berlin 1955, p. 101; Gabriel, op. cit., p. 51.

127 Lecture by Willy Hof at the Hafraba exhibition in Kassel, Kasseler Post 29 January 1928, UBK. On landscape aesthetics according to Todt, see Thomas Zeller 2002 and 2007.

128 Die Möglichkeit der Verwendung von Pflanzungen beim Bau neuzeitlicher Autostrassen, in: Verkehrstechnik, Heft 50, 1929, pp. 863–865.

6) Express highways abroad were discussed, especially those in the USA, Italy and Spain. For example, the Mount Vernon Memorial Park Road in the USA was featured.[129]

7) Reports were presented at international expressway and road construction conferences.[130]

Touring exhibitions in several cities were organized to promote the Hafraba project. In 1927, there were exhibitions in Cologne, Frankfurt, Hamburg and Hanover.[131] Moreover, there were exhibitions in the Trade Museum in Basel in 1927, in Kassel's Landesmuseum in 1928 and at Darmstadt's Trade Museum in 1928.[132]

Oddly, these exhibitions were only occasionally covered in the Hafraba newsletter. The Hafraba exhibition in Darmstadt was opened in the lobby of the Museum of Industry on 15 November 1928. The mayor, Dr. Glässing and the Darmstadt Chamber of Commerce and Industry extended the invitations. Chamber President Schenk emphasized the great interest of his chamber in the Hafraba approach and he acknowledged that by the end of the 1920s, the competition between motorized vehicles and the railroad would be a potential obstacle for the implementation on the Hafraba approach. Government Council Dr. Krebs gave a preliminary presentation about the Hafraba concept and emphasized the numerous initiatives in Italy for expressway construction. The timing was fortunate because the first "automobile only" road from Milan into the northern Italian lake district had just succeeded the 100-kilometre Milan–Bergamo–Brescia route.

---

129 Hafraba-Mitteilungsblatt, issue 5, 1932, p. 5. On the planning of the motorcars-only road Madrid–Irun with a total length of 360km, see Hafraba-Mitteilungsblatt issue 5, 1929, p. 4s.

130 Hafraba-Mitteilungsblatt issue 9, 1931, p. 1s contains the report on the first international Autobahn conference in Geneva from 31st of August till 2nd of September 1931; and the Hafraba-Mitteilungsblatt issue 5, 1932, p. 1s contains the report on the second international Autobahn conference in Milano from 18–20 April 1932. Furthermore: Bericht von internationalen Strassenbaukongress in Washington DC vom 6. bis 11. Oktober 1930, in: Hafraba-Mitteilungsblatt, issue 12, 1930, p. 12.

131 Kasseler Post on 29 January 1928, UBK.

132 Züricher Illustrierte Nr. 38, 1927 and Baseler Nachrichten on 31 August 1927, both Stadtarchiv Kassel, Akte Städtisches Verkehrsamt, Kasseler Post vom 10.2.1928, ibidem, Mitteilungsblatt issue 2, 1929, p. 7.

Image 13: Hafraba Exhibtion in the Town Hall of Frankfurt a.M. 1927.[133]

8.3 The Failure of the Hafraba

In the spring and summer of 1933, decisions and planning processes leading to the Nazi a*utobahn* project followed in rapid succession. At a general meeting of the Hafraba on 23 May 1933, its CEO Willy Hof reported on several conversations with Hitler about the Hafraba a*utobahn* concept. Hitler wanted to extend a network to cover the entire nation – a meeting on 6 April 1933 is documented in the chancellery files.[134] The editor of these files, Minuth, has stressed that there is no file to be found on the meeting between Hitler and Hof on 3 March 1933 as proclaimed by Kaftan – a meeting that Hof described in his report as "first reception and lecture".[135]

---

133 Hessisches Wirtschaftsarchiv Darmstadt.

134 Akten der Reichskanzlei, Regierung Hitler 1933–1938, ed. by Konrad Repgen, Teil I: 1933/34, Band 1, processed by Karl-Heinz Minuth, Boppard am Rhein 1983, p. 305, as well as report by the Hafraba manager Hof on his conversation with the Reich chancellor on the planning and financing of Autobahn construction on 6 April 1933, op. cit., p. 308.

135 Op. cit., p. 306.

On 24 June 1933, it was quickly decided to start constructing the Frankfurt–Darmstadt route. This route would be the first section of the Nazi a*utobahn* project. The common story is that this route was directly based on the plans of the Hafraba, although, in the archives, one can not find documentation to support this assumption.[136] Like in the Hafraba plan, there was a central strip between the lanes in opposite directions. Figure 11 shows the central strip of the a*utobahn* at Darmstadt in 1936.[137]

Figure 11: *Autobahn* at Darmstadt in 1936.

On 18 August 1933, the Hafraba association was compulsorily dissolved and incorporated into the project company Gezuvor – *Gesellschaft zur Vorbereitung des Reichs Autobahnbaus* (Company for the Preparation of *Autobahn* Construction).[138] Most of the original members of the executive board were included in the new company. The personnel records and portrait photos of the members were published in issue 9 of "*Die Autobahn*" in 1933.[139] The head office was transferred from Frankfurt (Main) to Berlin. The Gezuvor consisted of eleven regional sections with only regional responsibility. The Hannover–Kassel section was led by Kurt Becker. The closure of the Hafraba platform led to an increase of Todt's power. He kept

---

136 Völkischer Beobachter of 25 June 1933, p. 2, edition A. The Frankfurter Nachrichten of 30 June 1933 had the title "Ein Paradies für Autofahrer") on the beginning of the measuring works, Institut für Stadtgeschichte Frankfurt/Main.

137 Private picture by Siegfried Vahrenkamp, Osnabrück, 1936.

138 Die Autobahn, issue 9, 1933, p. 7–9.

139 From the old board of the Hafraba, the following were members of the NS party in August 1933: director Willy Hof, spokesman of the board Arthur Zierau, Niemeyer and Nienhaus, mayor of Heidelberg.

exclusive authority for planning the overall network and, in an 18 August 1933 speech before high-ranking state representatives on the occasion of the transformation of the Hafraba into the Gezuvor, he admonished the Gezuvor to focus on local issues and to refrain from central planning.[140]

It is noteworthy that the distinct visionary, and utopian character of the Hafraba *autobahn* project was transferred to the Nazi *autobahn* project and exaggerated into grotesqueness in the light of the low car density in Germany. This policy differed tremendously from the extension of the US highway network, which was expanded according to increasing needs following an exact cost-value-ratio, as proven by Seely.[141] According to the evaluation by Mark Rose, the US express highway system, which was built based on the German *autobahn*, didn't develop until the 1950s and was lagging behind the development of heavy overland traffic.[142] Mom shows that in the 1950s in Europe, an utopian approach of anticipating traffic volume before the fact was necessary in order to politically justify the construction of an *autobahn* network – he pointed out to Holland as an example of this.[143] The Nazi *autobahn* adopted the idea of collecting toll fees from the Hafraba as well. In §7 of the German Federal Law relating to the establishment of a business "*Reichs-Autobahngesellschaft*" (Corporation of the Reich Autobahn), dated June 27 1933, it is stated that tolls can be levied.[144]

Despite its many activities, the Hafraba association neither received much public attention nor did it gain acceptance. It's possible that Hafraba's propaganda push for the north-south route was unconvincing. As traffic counts conducted in 1924 and 1925 show, the main long-distance traffic flowed on the existing roads in the direction of Berlin–Cologne while north-south traffic was sparse.[145] This observation was also reflected in Nazi construction priorities. While the route Berlin–Cologne was finished by 1942, construction of the route Hanover–Hamburg was under construction. The idea of connecting Hamburg with Genoa recalls the

---

140 Städtisches Anzeigenblatt Frankfurt a.M. Of 26 August 1933, p. 378f, ISG.

141 Seely: Der amerikanische Blick auf die deutschen Autobahnen, op. cit.

142 Mark M. Rose: Interstate – Express Highway Politics 1939–1989, Tennessee UP, second rev. edition, 1990, p. 92.

143 Mom, op. cit.

144 Die Autobahn, issue 7, 1933, p. 1s.

145 F. Pflug: Der Kraftfahrzeugverkehr, in: B. Harms (ed.): Strukturwandlungen der Deutschen Volkswirtschaft, Berlin 1928, vol. 2, p. 251.

"geotechnical approach" of Fritz Stück, who projected "axes" without empirically proving the anticipated traffic.[146]

The federal association of the automobile industry was also opposed to the a*utobahn* plans of the Hafraba due to the high tax burden expected. However, the federal association totally misjudged the strategic opportunity of the a*utobahn* to create a market for automobile sales. In a letter to the chancellor on 6 January 1931, it argued that refinancing construction costs with the help of a toll would not work, as the Italian example had shown. Its letter responded to a memorandum of the federal government on "taxing motor vehicles"[147] which in section e) dealt with the question of expressway toll fees. The association rejected a toll since it would constitute an additional burden on car drivers who were already burdened with the raised fuel taxes of 1930.[148] The Federal Ministry of Finance presented this memorandum to the Reichstag in the early 1930s.

The attitude of the chambers of commerce and industry toward the private financing of expressways was not unanimous. The Essen Chamber of Commerce and Industry voted against the a*utobahn* project, opposing the six Hafraba members of the chamber. Already at the planning stage for the Bonn–Cologne a*utobahn*, Essen's Chamber of Commerce and Industry opposed a toll fee for this route – hinting at §13 – in a resolution submitted to the federal council (which was not at all responsible for these cases).[149] In his study, Thalmann shows that the chambers of commerce and industry of the Ruhr region did not have a common policy for a*utobahn* projects.[150] So far, the overall a*utobahn* policy of the chambers, has not yet been systematically explored.

The scant public attention paid to Hafraba's ideas can be verified by an evaluation of the business and traffic journals from 1930 to 1932. In most of the journals, the Hafraba project was not mentioned at all. Even *ADAC-Motorwelt* – the leading journal for motorists – makes

---

146 Folckert Lüken-Isberner: Fritz Stück und die Stadtplanungsdiskussion im Kassel der Weimarer Republik, in: Hessisches Jahrbuch für Landesgeschichte, Vol. 40, 1990, pp. 219–240.

147 Report of the Reichstag, Drucksache 620, 1930.

148 Institut für Stadtgeschichte Frankfurt/Main, Magistratsakten MA6289.

149 Report on the first meeting of the administrative council on Thursday, 10 February 1927, op. cit., p. 16–22.

150 Frauke Thalmann: Industrie- und Handelskammern im Ruhrgebiet während der Weimarer Republik, Berlin 1996, p. 250.

mention of the Hafraba project with only two short comments in the 156 editions that were evaluated for this study.[151] The *ADAC* backed the standpoint that the existing road network should be expanded. In 1931, on the occasion of a memorandum from the Federal Ministry of Finance on the taxation of motor vehicles, the *ADAC* demanded the allocation of tax revenue for the extension of the existing road network, and thereby, also the provision of jobs for the numerous unemployed.[152] Other automobile clubs followed suit. The Bavarian Automobile Club in Munich supported the Hafraba projects only half-heartedly: In a commentary that was published in the Hafraba magazine, it proclaimed the primary importance of the modernization of the existing road network already in existence over the new construction of an *autobahn*.[153] The Upper-Hessian Automobile Club in Giessen energetically opposed the Hafraba projects and sent a letter to the Chamber of Commerce of Giessen on 24 November 1928 dismissing the planned toll on *autobahns* and referring to a prior complaint from 5 June 1928 that the Automobile Club generally rejected the whole Hafraba project.[154] The Hafraba project would be untimely in regard to the current economic and road situation in Germany. In this respect, the thesis of Gijs Mom – that especially automobile clubs in Europe substantially supported Autobahn projects – must be revised.[155] This premise applies neither to the *ADAC* nor to the Upper-Hessian Automobile Club in Germany. Kurt Becker, member of the Hafraba board, regretted the disapproving attitude of the automobile associations in a speech delivered in Kassel on the occasion of the Hafraba exhibition.[156] Between 1926 and 1933, not a single Hafraba member contributed to the journal *Verkehrstechnik* which was a leading publication in the field of traffic policy until 1933. Rather it was the government officials of the federal states and provinces who published articles for it and ignored the Hafraba concept. Only once was there a small comment on the Hafraba project.[157]

The low impact of the Hafraba ideas can also be explained by the diverse interests of the automobile industry. Most German car engines were not built for the permanent stress of

---

151 ADAC-Motorwelt, 1931, Heft 25, p. 17–19, ADAC-Motorwelt, 1932, Nr. 19, p. 2–3.

152 ADAC-Motorwelt, 1931 Heft 25, p. 20–26.

153 Ludwig Graf Almeida: Kraftfahrer und Nur-Autostraße, in: Mitteilungsblatt Heft 11/1929, pp. 4–5.

154 Wirtschaftsarchiv Darmstadt 9/1749.

155 Mom, op. cit., p. 745s.

156 Kasseler Post of 29 January 1928, UBK.

157 Verkehrstechnik, 1928, p. 212.

driving on expressways and so that the German automobile industry had no good reason to supporting *autobahn* construction.[158]

[158] See A. Dinklage: Autobahnen verlangen andere Motoren, in: Die Autobahn, Heft 14, 1934, p. 638; and W. Kamm and H. Goll: Einfluß der Reichsautobahn auf die Gestaltung der Kraftfahrzeuge, in: Die Strasse, Heft 3, 1936, p. 67, further Edelmann, op. cit., p. 176.

9 The Autobahn Cologne–Bonn

An important milestone in the *autobahn* idea becoming reality was the junction-free Bonn–Cologne a*utobahn*. Its construction started in 1930 and it was opened on 6 August 1932 by Konrad Adenauer who was mayor of Cologne and also a member of the administrative council of the Hafraba – later he would be elected Chancellor of West Germany. The Hafraba reported in great detail on the opening in its magazine. This route was viewed as an entry into a new era.[159] The *autobahn* project Bonn–Cologne has to be evaluated in the context of traffic policy in the Rhineland Province of Prussia.

9.1 Transport Policy in the Rhineland Province

The Prussian Rhineland Province in the north spread out to the Lower Rhine (Cleves and Wessel), in the east across to Duisburg and Sterkrade, Essen, Elberfeld and Altenkirchen, and in the south to Kreuznach and Trier. Johannes Horion was elected governor of the Rhineland Province in 1922 and governed until 1933. The Rhineland Provincial Landtag held a meeting once a year in Düsseldorf. The Provincial Committee, of which Konrad Adenauer became the chairman in 1920, took care of the administrative work in between meetings and prepared presentations for the Provincial Landtag. In the Rhine Province Johannes Horion and Konrad Adenauer worked together and formulated the essential proposals for the annual Provincial Landtag. Horst Romeyk worked on the relations between Horion and Adenauer.[160] The key aspects of the Provincial Administration's work can be divided into the following three areas:

1. Socio-political tasks
2. Road construction and road traffic
3. Distribution of the motor vehicle tax to the community, according to the Prussian revenue sharing law and the Reich's revenue sharing law.

The Provincial Committee's traffic policy was strongly influenced by foreign policy – and especially sought to reduce the consequences of the Treaty of Versailles and the "Ruhr

---

159 Grossjohann (1932), pp. 1–5.
Hafraba Mitteilungsblatt 5 (1931), p. 6.

160 Romeyk 1976.

battle". Up to 1926 British troops had occupied the Cologne freight depot Eifeltor and curtailed train capacities.[161] According to the Treaty of Versailles, a four lane extension of the railroad line Cologne–Aachen was not permitted and this boosted the legitimacy of the expressway project Cologne–Aachen. Furthermore, the national borders led to the putting out of use for traffic of certain roads. This was the background for the provincial administration's proposal to take over the maintenance and administration of additional roads at the 69th Provincial Landtag in 1925. It was concluded that for „the Treaty of Versailles serious border area damage effects" a supplemental road network was required.[162] In 1923, when French troops occupied the Ruhr area and seized the railroad's carrying capacity, commercial vehicle traffic was able to fill the gaps in commercial transport,[163] but at the cost of deteriorated roads. These roads were not meant for truck traffic. After the Ruhr conflict ended, there was a high demand for road construction. As a consequence of this particular traffic situation, 20 percent of the Reich's commercial vehicles were located in the Rhineland in 1925.[164]

In the provincial committee's transport policy from 1925 to 1929, a wide range of big projects became evident. The focus was on the partaking of building for the Midland Canal, a high-speed railroad for the Cologne, Düsseldorf and Essen communities, the Nürburgring car racing track, the extension of communal roads into provincial roads and three different expressway projects. A comparable „project rush" on a provincial level – to which Cologne's mayor, Adenauer, had previously objected – became clear. On 31 March 1922 the Rhineland newspaper published – as an April fool's joke – a redraft of Adenauer's Cologne central station, which was supposed to be located on the hay market. An extremely modern monorail was going to be realized, which the Berlin entrepreneur Scherl brought into the debate for the future of the railroad.[165]

---

161 Pohl 1933, p. 37.

162 Verhandlungen des 69. Rheinischen Provinziallandtages, Juni 1925, p. 94.

163 Cohausz 1973. Stellungnahme der Industrie- und Handelskammer Köln zum Autobahnplan Köln–Düsseldorf vom 22.10.1926, in: Historisches Archiv Köln (im Folgenden: HAStK), Akte Adenauer 902, Nr. 68, Fasz. 22-2.

164 Reifferscheidt 1926, p. 46.

165 HAStK, Akte Adenauer 902, Nr. 82, Fasz.1. Scherl 1909. For a review of Scherl's concept see Zeitschrift des Vereins deutscher Ingenieure, 1909, p. 1038.

Image 14: The new Cologne Main Station as a Joke.[166]

For the 69th Provincial Landtag in 1925, a proposal was presented on the participation of the Rhine Province in the planned Rhenish-Westphalian High-Speed Railroad AG. The high-speed railroad project had already been discussed several times since 1909.[167] The goal was to build a new railroad line from Cologne to Dortmund, which was supposed to meet the modern requirements of passenger service: A frequent succession of trains, a fixed schedule, and high speed. At the time, the *Reichsbahn* was not able to meet these quality criteria since it provided mixed passenger train and freight train traffic on the line, as well as mixed traffic from local and long-distance trains. The representative Schäfer affirmed that years of work would have to be carried out before the concession for a high-speed railroad could be issued. People concerned about mining opposed this project because it was feared that there would be mine subsidence damage. In the end the high-speed railroad project failed.[168]

The records of Provincial Landtag's negotiations for road administration work were very well documented with maps (cf. Image 15) and statistics for traffic and developing road conditions. The province administered 6345 km of roads for which it hired 800 road surveyors.[169]

---

166 HAStK, Bestand 1010, Nr. 34, press cuttings.

167 Quatram 1926.

168 Verhandlungen des 69. Rheinischen Provinziallandtages, Juni 1925, p. 28.

169 Bericht des Provinzialausschusses der Rheinprovinz über die Ergebnisse der Provinzialverwaltung für das Regierungsjahr 1. April 1924 bis 31. März 1925, p. 16.

The provincial administration's road building policy goal was to improve the quality of roads in order to meet the demands for growing motorized transportation. Dangerous railroad crossings were to be made safer which indicates that the system of the emerging motorized traffic was not sufficiently decoupled from the railroad. Another goal was to provide roads with a concrete coating to prepare for the major burdens from the motorized transportation. Experiments were made with basalt coatings and tar coatings. Since the basalt quarries were overburdened by 1924, one turned to technology for the tar coating.[170] In later years this situation was reversed and additional orders for basalt coating were made to provide jobs for the depressed basalt quarries.[171]

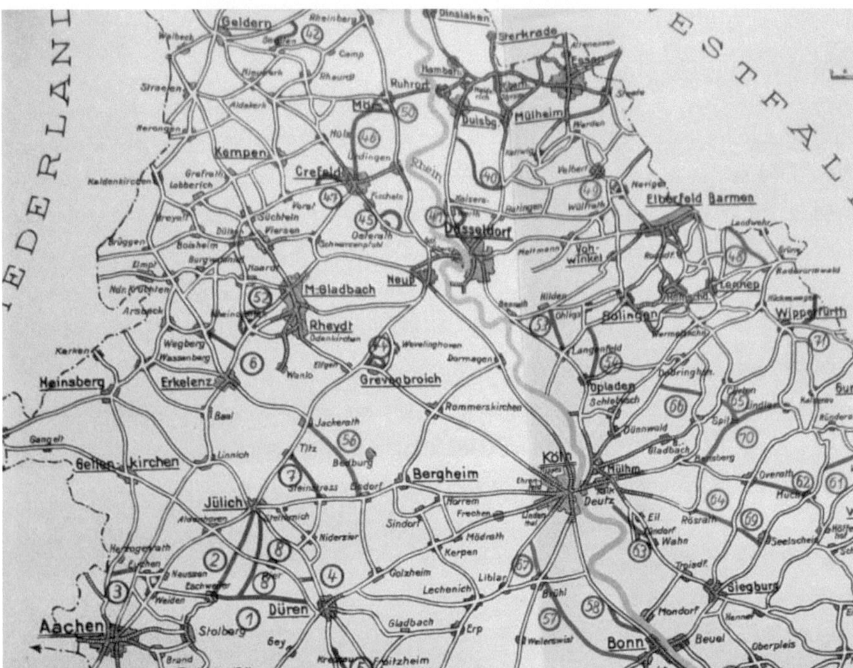

Image 15: Clipping of the Road Network in the Rhine Province[172]

---

170 Bericht des Provinzialausschusses der Rheinprovinz über die Ergebnisse der Provinzialverwaltung für das Regierungsjahr 1. April 1924 bis 31. März 1925, p. 17.

171 Bericht des Provinzialausschusses über den Fortgang des Ausbaus der Provinzialstrassen im Jahre 1926, in: Verhandlungen des 73. Rheinischen Provinziallandtages April 1927, Drucksache 22, p. 81s.

172 Verhandlungen des 73. Rheinischen Provinziallandtages April 1927, Anlage 25, p. 82.

From 1925 to 1929 one can observe that the provincial committee's transport policy had continuously increased its tasks. An important agenda item in the Provincial Landtag's proceedings was the taking over of roads from local administration and provincial county administration. In 1925 the 69th Provincial Landtag decided on a „future program" of gradually conveying 2000 km of local roads to the Provincial Administration. According to a progress report from 1929, 410 km had been taken over by that year. The takeover was intended specifically for the maintenance of the road scene, and for the maintenance and extension of roads according to provincial standards – and for the application of a surface coating.[173]

In the provincial committee's report and proposal for the Provincial Administration's takeover of additional roads for maintenance and administration to the 69th Provincial Landtag from the 12th to the 17th of June, 1925, the provincial committee realized a change of meaning for country roads. In the 19th century the railroad had diminished the significance of country roads since these roads were mostly used for local transportation to a train station. But in the years that passed, the automobile was responsible for an increase of importance for roads: „Today the wholesaler serves his customers in a perimeter of 100 kilometers directly from his commercial vehicle all the way to the most secluded places. In many cases the commercial vehicles led to cheaper – and above all – faster service. Long distance driving without frequent loading and unloading resulted. The number of automobiles vastly increased and became an essential means of transportation for business"[174]. For this reason, the demand for roads suitable for modern vehicles increased.

In order to advance road building policy in the Rhineland, transport policy issues were skillfully related to social policy questions. In Johannes Horion's budget speech before the Provincial Landtag, he first withdrew from the overwhelming unemployment problems. Later he pointed out how to increase employment through public contracts for road construction. A

---

173 Bericht des Provinzialausschusses über den Stand des Ausbaus und die Übernahme weiterer Strassen in die Unterhaltung und Verwaltung des Provinzialverbandes, in: Verhandlungen des 75. Rheinischen Provinziallandtages April 1929, Drucksache 29, p. 140.

174 Bericht des Provinzialausschusses betreffend die Übernahme weiterer Strassen in die Unterhaltung und Verwaltung des Provinzialverbandes, in: Verhandlungen des 69. Rheinischen Provinziallandtages April 1925, Drucksache 23, p. 92.

large part of the financing came from productive social welfare funds. The other part was financed by loans. The Rhine Province accrued huge extraordinary household expenses with its road building policy – these were financed by borrowing. In his budget speech in front of the 75th Provincial Landtag in 1929, Horion indicated the risks from huge borrowing, but stated that the Rhine Province's strong revenue increase from the motor vehicle tax could cover these risks.[175]

In addition to the general road building policy, the provincial committee had to work out a proposal to support building the Nürburgring race track[176] and needed three proposals to support building expressway routes. The routes Aachen–Düren–Cologne, Cologne–Düsseldorf and Cologne–Bonn were at issue. Recommendations for building these expressways were presented in the Provincial Landtag's proposals. The Provincial Landtag supported and approved the expressways – but only the Bonn–Cologne *autobahn* project went into the phase of realization. Since the powerful *Reichsbahn* regional administration had its headquarter in Cologne, the provincial committee's expressway plan was directly administered by the *Reichsbahn* – leading to corresponding backlashes.

As with the Rhineland's general road building policy, transport policy questions for the expressway projects were also related to social policy questions. A major part of financing was supposed to come from productive social welfare funds. The other part, which had to be financed by loans, was to be paid off later by tolls once the expressway went into operation. The concept was quite convincing and conclusive, and therefore it was widely supported by the Rhineland Provincial Landtag. In order to levy the crucial expressway tolls for the financing, a certificate of exemption was required from the Reich government in Berlin. This was due to §13 of the the *Finanzausgleichgesetz* (revenue sharing law) which forbade tolls on public roads. However, the Rhine Province's concept to refinance the expressway with tolls was unsuccessful. And so the expressway Bonn–Cologne had to be financed solely by productive social welfare funds and loans without the guarantee of a future back-flow of funds.

---

175 Stenographische Berichte über die Verhandlungen des 75. Rheinischen Provinziallandtages März 1929, p. 16.

176 Bericht des Provinzialausschusses betreffend Unterstützung der Autorennstrasse „Nürburg-Ring“ (Kreis Adenau), in: Verhandlungen des 71. Rheinischen Provinziallandtages März 1926, Drucksache 27.

The proposals for building the expressway routes contain various arguments for the need for the expressway construction given the increasing traffic. They included concepts for planning routes with extensive map materials, statements of costs incurred, and financing plans. As with the proposals to the Provincial Landtag for general road building policy, there are three proposals for expressway construction – each signed by Adenauer, the chairman of the provincial committee and by Horion, the governor. With this, Adenauer's prominent role in planning the first expressways is indisputably documented. In the proposals, the improvement of traffic conditions is not the only goal. The second goal, to create jobs, was a part of the social policy ideas of Adenauer and Horion, both members of the Centrum Party. Providing public relief work to relieve high unemployment was also devised by Adenauer as the mayor of Cologne.[177]

In their proposals, the Provincial Administration commenced with an observation about the strong increase of car traffic and the special performance of automobile technology for cars and trucks. The Provincial Administration also indicated that this technology was superior to the railroad technology. It also pointed out – in its proposal for the expressway Aachen–Cologne – the railroad's poor performance in commercial transport as a reason to build the road. Of course, with these policies the Provincial Administration evoked the distrust of *Reichsbahn* administration in Cologne which subsequently enforced a strictly negative position towards expressway projects at the Ministry of Transport in Berlin.

9.2 The Autobahn Project Aachen–Düren–Cologne

Each of the three expressway projects developed very differently. The project Aachen–Düren–Cologne was formulated in June 1925.[178] Through an improved transportation infrastructure, the goal was to relieve truck traffic and therewith to support manufacturing Industries in the Düren and Aachen areas. In the proposal it was explained that the facilitation of transport for industrial packaged goods traffic was vital, since the rail line Aachen–Cologne was not adequate for the amount of demanded traffic. In 1922, Aachen rail freight depots had

---

177 Schwarz 1986, p. 199, p. 291. Frielingsdorf 2002.

178 Bericht des Provinzialausschusses betreffend den Bau einer Autostrasse Aachen–Köln, in: Verhandlungen des 69. Rheinischen Provinziallandtages 1925, Drucksache 24, pp. 100–103.

to close on 170 out of 300 workdays because of overloading, and, according to the Treaty of Versailles, a four-track extension for the railroad from Aachen to Cologne could not be built.

The criticism against the *Reichsbahn* led to an instant reaction from Cologne's *Reichsbahn* administration. The day before a scheduled Provincial Landtag meeting on 12 June 1925, a press statement from the railway administration appeared in the Cologne newspaper. The *Reichsbahn* statement claimed that, contrary to the provincial committee's statements, in 1922 they closed the stations for commercial transport for only six days. The *Reichsbahn,* it insisted, was quite productive and could increase transportation for the route Aachen–Cologne by 40 percent. This press maneuver by the *Reichsbahn* led to several irritations in the Provincial Landtag since the representative Schäfer referred to this press report in plenum to challenge the credibility of the provincial committee's proposal.

The driving conditions, which were described as unsustainable, threatened the core of industrial existence according to the proposal, and needed to be improved. It was explained in the proposal that: „A road would be beneficial for the greater Aachen–Düren region's packaged-goods-traffic and especially for the finished goods manufacturing such as the textile, paper, needle, and glass industries".[179] These statements are remarkable for introducing extraordinarily innovative traffic and logistics concepts that suggest that truck traffic and the general cargo transport infrastructure are related to manufacturing and the expressway. This concept radically dispensed with the prevalent rail logistics at that time. In the 20th century this traffic concept became trend-setting for the development of expressway infrastructure throughout Europe – for truck traffic, for modern logistics and for the differentiation of the manufacturing production stages, as well as their locations.[180] Indeed this traffic concept developed in the following years, without support from German transport policy at a state or federal level, since the state and national railroads were able to assert their interests in this political field – against the commercial vehicles – until the 1980s.

In the proposal for the expressway Aachen–Düren–Cologne to the 69th Rhineland Provincial Landtag, it was also emphasized that the project was of greatest interest for the workforce

---

179 Ibidem, p. 101.

180 Vahrenkamp 2009, Vahrenkamp 2010.

with its high rate of unemployment. Its financing concept was the basis of the project. Funds from the productive unemployment aid were to be used to cover one third of the cost and the remaining two thirds were to be raised by the participating local organizations which, in turn, also had to shoulder two thirds of any possible deficit once the expressways went into operation. A 5 Pfennig/kilometer toll was to be levied per car (10 *Pfennig* for a truck). In March of 1926, the provincial committee concluded that the public financial system had deteriorated and that the participating local organizations could not meet the financial requirements. This was especially the case as the expected toll revenue from the expressway in operation would not be able to raise the necessary amount owing to the limited traffic. The project had to be suspended.[181]

In establishing the project in Berlin, Adenauer's contacts in Berlin initially appeared to be very helpful in achieving his goals. Apart from his major function as mayor and his function as the chairman of the provincial committee, he was also the acting chairman of the Prussian State Council. To meet his obligations it was incumbent on Adenauer to spend two to five days out of every month in Berlin.[182] At the beginning of July 1925, Horion went to Berlin in order to provide support for the expressway project Aachen–Cologne and, to his surprise, discovered that about ten Reich ministries or other central departments participated in the decision-making for the project.[183] How strongly both Horion and Adenauer pulled together the road project becomes clear to Adenauer in Horion's letter on 30 June 1925, where he says about his discussions in Berlin: „It would naturally be extremely beneficial if you throw your heavy weight onto the scale and participate in the discussion".[184] However, reports on the negotiations in the Welfare Ministry reveal that Adenauer's influence in Berlin was limited despite his frequent stays there and he was hardly able to positively influence the negotiations in the Prussian ministries.[185]

---

181 Bericht des Provinzialausschusses: Die Zunahme des Verkehrs auf den Provinzialstrassen, in: Verhandlungen des 71. Rheinischen Provinziallandtages 1926.

182 Adenauer 1976.

183 Horion letter to Adenauer 30 June 1925, HAStK, Akte Adenauer 902, Nr. 82, Fasz.1.

184 Ibidem.

185 Geheimes Staatsarchiv Preußischer Kulturbesitz (in the following: GStA PK), HA Rep. 151 Finanzministerium, IC Nr. 11939.

9.3 The Autobahn Project Cologne–Düsseldorf

The second project, Cologne–Düsseldorf, began in 1926. In agreement with the communities, the Rhineland Provincial Administration came up with the draft for the route Cologne–Düsseldorf. In order to set the stage for a new expressway project right after the aborted Aachen–Cologne project, the proposal to the Provincial Landtag highlighted that the new project would be more cost-effective because of the heavy traffic expected and the concomitant high toll revenues.[186] For the toll, once again 5 Pfennig per kilometer would be levied per car and 10 Pfennig per truck. The project was justified by pointing to the heavy traffic on the Cologne–Düsseldorf route and the strong expected increase, as well as by high unemployment in the Cologne–Düsseldorf area. In fact, the project would become widely supported. It was, however, criticized by deputy Theißen. He viewed the expressway Cologne–Duisburg as competion to the planned high-speed railroad and urged a vote on both concepts.[187]

The proposal to the 72th Provincial Landtag, which met at the end of December 1926 in an extraordinary proceeding, linked the project explicitly to Cologne's decision to build a Rhine bridge near Mülheim. This was considered to be an additional justification for the project. The literature points out numerous city development projects that were managed by Adenauer as the mayor of Cologne.[188] Now the connection between both of Adenauer's project levels – in Cologne and the Rhine Province – can be recognized.

The provincial committee's proposal was accompanied by an exposé of the expressway Cologne–Düsseldorf which Horion had prepared on 1 December 1926 and in which he delineated the essential features of the project.[189] Image 16 shows a map of the projected expressway. Initially Horion showed interest in the exposé on increasing motorized traffic in different countries and also deduced from it a strong increase in Germany for the near future. Compared to other countries, Germany had a lower per capita motorized vehicle density.

---

186 Bericht des Provinzialausschusses betreffend den Bau einer road Köln–Düsseldorf, in: Verhandlungen des 72. Rheinischen Provinziallandtages Dezember 1926, Drucksache 6, pp. 6–8.

187 Stenographische Berichte über die Verhandlungen des 72. Rheinischen Provinziallandtages Dezember 1926, p. 186.

188 Schwarz 1986. Frielingsdorf, 2002. For the bridge Mülheim see Frielingsdorf 2002, p. 108–110.

189 HAStK, Akte Adenauer 902, Nr. 68, Fasz. 22-2.

Then he responded to the traffic census of the existing road Cologne–Düsseldorf in 1925, which demonstrated average daily traffic of 1500 vehicles per day, of which one third was trucks. The Bonn–Cologne road was among the most heavily trafficked roads in the Rhine Province. The exposé assumed a 100 percent traffic increase for the road from 1925 to 1930. Because of its major importance for commercial vehicle traffic, the *Reichsbahn* turned towards this business area. It operated a public transport system itself in the Düsseldorf area with its commercial vehicles – on the one hand to counter the private commercial traffic, and on the other hand to regain as much of the traffic that had been lost to the country road.[190] The alternative, to build bypasses in order to save communities from through-traffic, was tested, but discarded. In Horion's exposé he visualizes for the expressway two lanes per driving direction of 4 meters each resulting in a roadway with a total width of 12 meters. Additionally, he envisioned 2-meter shoulders in each direction. A four-lane extension was projected based on the lessons learned from the expressway in Italy. On the route between Cologne and Düsseldorf, seven exits were planned; these were denoted as car stations.

On 24 August 1926, the governor of the Rhine Province forwarded Horion's report to the Prussian Minister for the Nation's Welfare in Berlin together with an extensive draft for the expressway Düsseldorf–Cologne and requested, „by law §19 the regulations are given for public relief works on 30 April, 1925 by the Reich labor administration, for intensified support through granting a 10 year loan, which is five times the amount of the saved unemployment support."[191] Consequently, the Minister for the Nation's Welfare organized a „departmental meeting" on 4 October, „for the purpose of formulating a consistent Prussian opinion on the Rhine Province's plan to build the expressway Düsseldorf–Cologne".

---

[190] Derboven 1929, p. 17.

[191] GStA PK, HA Rep. 151 Finanzministerium, IC Nr. 11939.

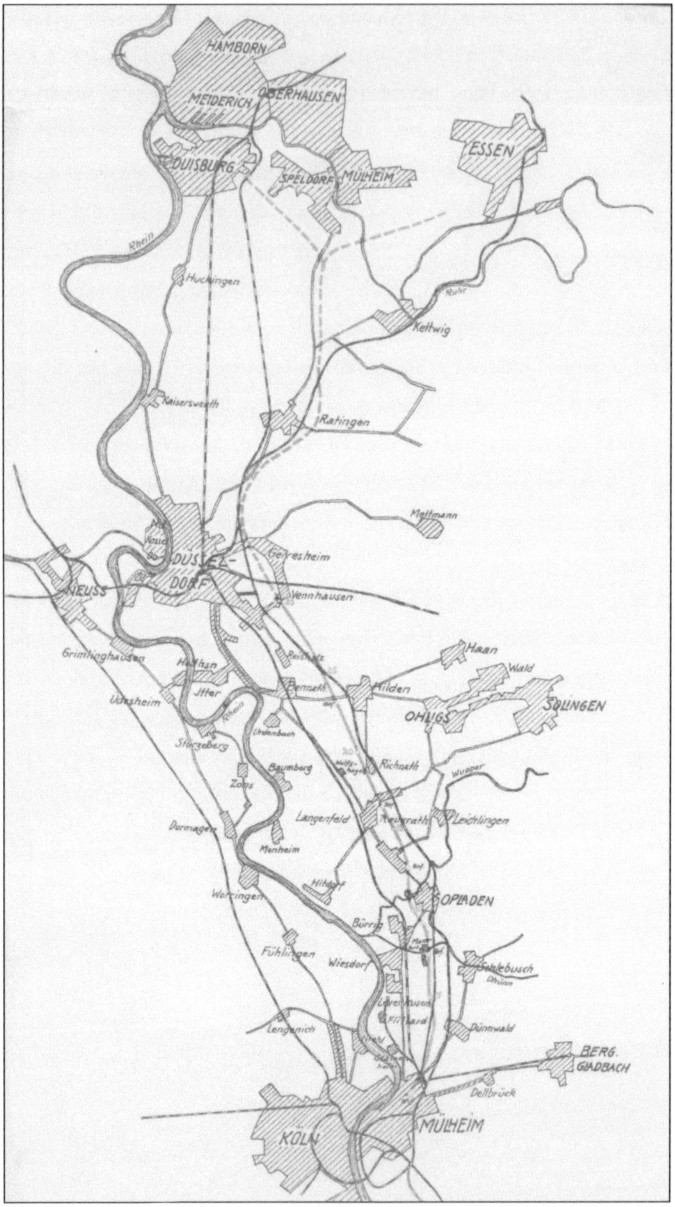

Image 16: Autobahn Cologne–Düsseldorf–Essen (dotted line) in the Memoir of Horion

Horion's expressway project initially faced widespread disapproval at the departmental meeting in Berlin.[192] The representatives of the Ministry of Agriculture expressed doubt that, based on the present law, it would be possible to levy a toll for the use of the road. They asked for proof that road building in the rest of the Rhine Province would not be impaired. The representative of the Ministry of Trade explained that carrying out expropriations would be feasible only if the vital necessity for building the road was proven. The representative of the Ministry of Finance put forward financial doubts and requested an inquiry on whether the Provincial Administration would also be content with an interest subsidy for several years. The various involved parties had valid doubts regarding the expected cost-effectiveness of the project.

In the end, the district presidents of Cologne and Düsseldorf, who took part in the meeting on 4 October 1926, stabbed Horion in the back. The district president of Cologne realized that the traffic conditions on the provincial Cologne-area road to Düsseldorf were satisfactory. For this reason there was no public interest in carrying out expropriations for land acquisition. The district president in Düsseldorf also described the present traffic in the towns Benrad, Langenfeld, Opladen and Wiesdorf as reasonable, so that the support for construction of an expressway was completely unnecessary.

The objection of the Düsseldorf district president was surprising since on 28 September 1926, the district administrators of the Düsseldorf County Council and the Opladen county – probably under Horion's orders – sent resolutions to the district president in which they pointed out the unsustainable conditions on the Cologne–Düsseldorf road. The District Administrator of the Düsseldorf County Council, von Chamier, explained that the county council's unanimous decision was at issue to support, „in the interest of the provision of employment and safe traffic conditions, the immediate push to build the road". The *Reichsbahn's* influence is mentioned: „With great surprise the county council noticed that the immediate initiation of this project, which was ready for building, would jilt the *Reichsbahn* administration's pleas to address the road as a competing business."[193]

---

[192] Note of Horion about a meeting in Berlin on 4 October 1926 in Berlin, in: HAStK, Akte Adenauer 902, Nr. 82, Fasz. 1.

[193] HAStK, Akte Adenauer 902, Nr. 82, Fasz. 1 and GSA HA Rep. 151 Finanzministerium, IC Nr. 11939.

The Prussian Ministry of the Nation's Welfare requested an assessment of the project by the Chamber of Commerce and Industry of Cologne, which had emphasized the need for the expressway in a report on 29 October 1926.[194] To prepare the report, Horion invited Adenauer to a meeting on 15 October 1926 at the chamber in Cologne.[195] The Düsseldorf Chamber of Commerce and Industry had also given a positive evaluation of the project[196] and induced the district presidents of Cologne and Düsseldorf to change their policy: now they supported the project and the district president of Cologne acknowledged public interest in the expropriation of land.[197]

The positive report by Cologne's Chamber of Commerce and Industry was well-received in Berlin. On 10 November 1926, a departmental meeting in the Prussian Ministry for the Nation's Welfare took place and results were forwarded to the Prussian Finance Minister on 13 November. The economic need for the road was recognized. With regard to the consideration of levying a toll, Martini, the councilor for the Minister of the Interior, objected „that the expressway Cologne–Düsseldorf should not be seen as an independent transport facility in connection with §13 of the Reich's revenue sharing law. It was left open whether or not the expressway financing with a toll would be in accordance to the §13 revenue sharing law (which did not allow a toll on public roads).[198] Hence levying a toll was by no means possible. On the other hand, the same paragraph provided the possibility of granting exemptions through the State Council (Reichsrat). The case at issue was suitable for a corresponding proposal to the State Council. It was generally agreed that the business was eligible for assistance. But since the initiation of assistance is dependent on settling the question of toll fees, a proposal to the State Council which would be submitted by the Welfare Ministry – via the Ministry of State – must first await the requested opinion of the Interior Minister.[199]

---

194 HAStK, Akte Adenauer 902, Nr. 68, Fasz. 22-2.

195 HAStK, Akte Adenauer 902, Nr. 82, Fasz. 1.

196 GStA PK, attachment of the letter of Preuss. Ministers für Volkswohlfahrt to den Preuss. Finanzminister, 13 November 1926, HA Rep. 151 Finanzministerium, IC Nr. 11939.

197 GStA PK, Der Regierungspräsident Köln an den Preussischen Minister für Volkswohlfahrt, 6. November 1926, HA Rep. 151 Finanzministerium, IC Nr. 11939.

198 Finanzausgleichsgesetzt vom 9 April.1927, Reichsgesetzblatt I, 1927, p. 91.

199 GStA PK, HA Rep. 151 Finanzministerium, IC Nr. 11939.

In Horion’s budget speech in front of the 72nd Provincial Landtag on 29 December 1926, he described the situation in Berlin as promising.[200] He had been able to get backing from the five participating Prussian Ministries for the project. He explained that the Reich Ministry of Labor and the Reich Labor Administration had agreed on issues concerning the making available of productive unemployment aid and might have persuaded the Reich Ministry of Finance to take this position as well. However, the Reich Ministry of Transport was non-committal. It was currently in negotiations with the *Reichsbahn*, which was negatively inclined because of the competition that the expressways posed. The interesting part of Horion's testimony is obviously the powerful statement about the Ministry of Transport as being compelled to negotiate with the *Reichsbahn*. Furthermore, a resolution of the State Council had to be made to legitimize the toll by approving an exemption from §13 of the revenue sharing law. A corresponding Prussian proposal already existed at the State Council.

Resistance to the expressway project came from different quarters. The ministries feared launching a precedent. A toll charge for the Mannheim–Heidelberg expressway project could not be denyed if it was allowed for the Cologne–Düsseldorf route. One was generally against a toll since it would be akin to a “medieval road toll”.

The powerful *Reichsbahn* was against the project. In a report to the Prussian Finance Minister on 23 December 1926, the Director-General Julius Dorpmüller, requested that the *Reichsbahn* oppose the plan. Construction activities, he concluded, would serve special interests at the „expense of the community and its traffic could only be won at the expense of the existing – not full exploited – railroads of the *Reichsbahn*. Such an expensive construction can be justified only when the costs are borne by the interested parties.” A report to the Chamber of Commerce in Stolberg suggested that the anticipated toll revenues were exaggerated and that the „cost of depreciation“ may not even be covered by the levying of a toll. As a result, it was expected that the expenditures would not be covered by the revenues.[201]

---

200 Stenographische Berichte über die Verhandlungen des 72. Rheinischen Provinziallandtages March 1926, p. 9.

201 GStA PK, HA Rep. 151 Finanzministerium, IC Nr. 11939.

Horion was asked to comment on the Reichbahn's accusations . In Horion's written reply to the Prussian Finance Minister on 25 January 1927, he refutes the accusations: „The state railway's antagonism to good roads with the reasoning that good roads are not necessary because nobody needs to drive a car since everyone could just as well use the train, will probably not be supported also by other agencies than those being obligated to maintain the roads."[202]

In particular, tensions developed between the settlement organization Ruhrkohle and the provincial committee. The settlement organization in the Rhineland Province was responsible for road administration in the Essen and Oberhausen regions. On the organization's board were two members from Stufa[203] – an association of road planners that rejected the expressway in favor of expenditures for the extension of the existing road network. This position was also shared by the leading ministry officials of the Federal States of Bavaria and Saxony – Vilbig and Speck. The controversy in the transport policy of the 1920s, thus, centered on the options of new expressway construction versus the extension of existing country roads. In his budget speech to the 72nd Provincial Committee on 29 December 1926, Horion explained that the Rhine settlement organization had been induced by „Berlin officials" to reject the planned expressway project in its report. The Rhine settlement organization had even distributed the report to the members of the provincial committee. Here one can see tensions between the raw material industries in the region of Essen and the finished goods industries in the region of Düsseldorf. Similarly, the automobile club ADAC rejected the new expressway construction and supported the position of Stufa.

The Chamber of Commerce in Essen was also against the expressway project and intervened at the State Council, where a decision would be felled on the proposal to grant an exemption from §13. In a report on the 20 January 1926 the chamber explained: „Since the decision on whether or not to allow a toll charge is in the State Council, we request of the gentlemen members of the State Council to take into account that we have an overall negative opinion of the project for the following reason ... The Chamber's view was that the existing roads could be improved and bypasses built with significantly lower funds. Building a special expressway

---

[202] Ibidem.

[203] Die Studiengesellschaft für Automobilstrassenbau. On Stufa also see Gabriel.

should be postponed to a future date ... Given the unfavorable economic climate, an expenditures of 40 million RM for the expressway Cologne–Düsseldorf and its extension to Duisburg–Essen, was not warranted. In any event, it is more appropriate to follow the Rhine settlement organization's plans for a maximum network of roads to serve car traffic. As far as the introduction of fees is concerned, scepticism is in order because such tolls are indistinguishable from the medieval road tolls and can choke off traffic. Car traffic is already burdened by the motor vehicle tax fees and the bridge toll levy."[204]

At the 73rd Provincial Landtag in April 1927, in the run up to the expressway project Cologne–Düsseldorf, it was brought up that the district president in Düsseldorf had detailed plans to be examined. The tender documents had been prepared so that the tender and allocation would be possible as soon as ownership of the building site was guaranteed.[205] All preparations had been made for land acquisition to begin right after the State Council's approval of the toll inquiry. The total costs were calculated at 17 million RM, of which 2,2 million was for land acquisition, 4,0 million for earthworks, 2,3 million for construction and 5,6 million for the roadway.

In Horion's budget speech in front of the 73rd Rhineland Provincial Committee on 4 April, 1927, he gave a pessimistic assessment for the future prospects of the project.[206] He referred to a recent meeting in the different Berlin ministries and gratefully acknowledged the representatives of the Prussian state government, from which he learned about the project. He also received support from the representatives of the Reich Ministry for Occupied Areas and from the representative of the Rhine Province in the State Council. At one point, though, he started to get the impression that the Reich authorities needed to focus on the difficulties and misgivings about the project and the prospects for completing it dwindled. At the state level, now all the ministries began to voice doubts. The Ministry of Finance agreed to release the funds for the productive unemployment aid for one year only and it now became incumbent

---

204 GStA PK, HA Rep. 151 Finanzministerium, IC Nr. 11939.

205 Verhandlungen des 73. Provinziallandtag: Bericht des Provinzialausschusses betreffend den Stand der Vorarbeiten, der Voranschlag und die verschiedenen Anträge der KPD bezüglich Entlohnung der Arbeiter beim Bau der Autobahnbau Köln–Düsseldorf, Anlage 37, pp.128–131.

206 Stenographische Berichte über die Verhandlungen des 73. Rheinischen Provinziallandtages März 1927, p. 16s.

on the Provincial Administration to prove that it was able to raise 4 million RM loan. The Provincial Administration also had to demonstrate that, in the case that the levy toll was not approved, that it could assume the interest payments and amortization rate. Since the project had already been committed to paper in Berlin for six months, Horion realized that two things were amiss by the crucial ministries; an interest for the accelerated provision of work for the unemployed and for the improvement of traffic conditions.

Horion's assessment is supported by the records of the Prussian Welfare Ministry according to which the government of the Reich refused to deliver a proposal to the State Council for the exemption from §13. The Prussian proposal was denied by a cabinet decision on 17 June 1927, but generous funding for the project was agreed on.[207] The Reich government and the State Council's interpretation of the law prohibited a toll charge for public roads. The Rhineland Province would have to operate the planned expressway as a private road to levy a toll for it. For this solution, legal difficulties are identified in two areas: As a private road, no public interest for the expropriation of land could be expected. And car drivers could be successful at taking legal action against the toll. The financial basis of the toll charge was thus at risk.

9.4 The Autobahn Project Cologne–Bonn

Because of these difficulties, the expressway project Cologne–Düsseldorf was delayed and, in 1929, the expressway project Cologne–Bonn was proposed. The latter was to be understood as a southern extension of the delayed Cologne–Düsseldorf project. Once again, §13 of the *Finanzausgleichsgesetz* (revenue sharing law) did not allow a levying of tolls on public roads and Adenauer and Horion's project of did not get support at the federal government level. An exemption from §13 could not be achieved although it could have been legislated. As a

[207] In the protocol of the cabinett meeting on 17 June 1927: „Die Verbesserung der Verkehrsverhältnisse, wie sie durch die Erbauung einer Autobahnstrasse von Köln nach Düsseldorf seitens der Rheinprovinz erstrebt wird, findet die Billigung und Unterstützung der Reichsregierung; sie ist bereit, einen verlorenen Zuschuß von 3 Millionen und weitere 10 Millionen Darlehn aus der produktiven Erwerbslosenfürsorge zur Verfügung zu stellen. Die Reichsregierung bedauert aber, dem preussischen proposal auf Zulassung der Erhebung von Abgaben für die expressway Köln nach Düsseldorf im Hinblick darauf, dass es sich um einen öffentlichen Weg handelt, nicht zustimmen zu können. Damit bleibt selbstverständlich die Frage unberührt, ob die Angelegenheit durch Schaffung einer Privatstrasse mit Benutzungsentschädigung geregelt werden kann." Die Historische Kommission bei der Bayerischen Akademie der Wissenschaften (eds.): Akten der Reichskanzlei Online, Das Kabinett Marx III/IV.

compromise, the Bonn–Cologne Autobahn was given the go-ahead, but under an alternative financial scheme; the construction was to be covered by the federal government's fund for emergency works.

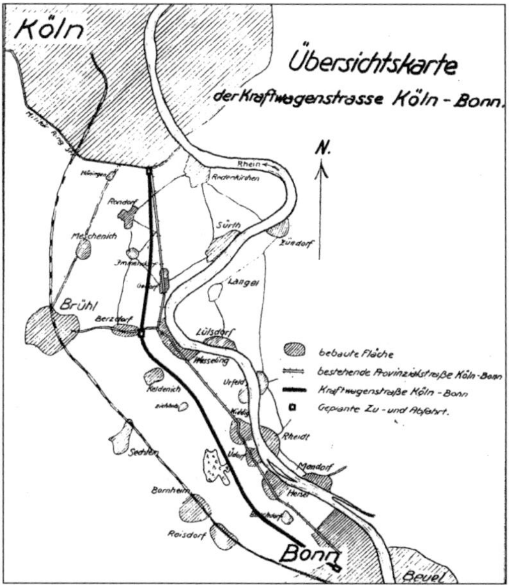

Image 17: Autobahn Bonn–Cologne

For the building of the expressway Bonn–Cologne, 11 million RM were budgeted – partially from productive unemployment aid funds. Instead of using heavy equipment, the necessary soil removal was realized by manual labor to secure more jobs for the numerous unemployed. Five-and-a-half thousand previously unemployed men were to handle ca. 300.000 daily tasks[208] so that the goal could be achieved. Image 17 shows the course of the *autobahn* Bonn–Cologne.[209]

---

208 Bonner General Anzeiger on 6 August 1932, Archiv des Landschaftsverband Rheinland.

209 ADAC-Motorwelt, 14 February 1930, p. 27.

Image 18: Cover page of the ADAC-Motorwelt on 25 July 1932

Konrad Adenauer, the mayor of Cologne, declared the route Cologne–Bonn as "open to traffic" in a ceremonial act in the Cologne town hall on 6 August 1932. In his speech, he pointed out the road's futuristic concept: "That is what the roads of the future will look like.“[210] The opening of the expressway was celebrated by the *Allgemeine Deutsche Automobil Club (ADAC)* with a rally from locations all over Germany to Cologne on 7 August 1932. The

[210] Grossjohann: Die Kraftwagenstrasse Bonn–Köln–Düsseldorf, in: Strassenbau und Strassenunterhaltung, Beilage der Zeitschrift Verkehrstechnik, 1929, p. 415–417, also: Gabriel, op. cit., p. 40–42. Kölner Stadtanzeiger on 1 August 2007, Jürgen Hahn: 50 Jahre Autobahnen in Deutschland, in: Strasse und Autobahn, Heft 1, 1983, pp. 1–4.

Hafraba reported extensively in its magazine about the inauguration.[211] The *ADAC-Motorwelt* promoted this road in its title page of the 25 July 1932 issue which is shown in Image 18. In the interesting layout, tradition (Cologne Dome) is juxtaposed with modernity. Apparently the advertising director assumed that the recognition of new roads could be improved by relating them to tradition. Horion edited a booklet on the occasion of the opening of the first German *autobahn.*[212]

Image 19: Opening of the Autobahn Bonn–Cologne on 6 August 1932.[213]

The Bonn–Cologne *autobahn* was intended for 'automobile use only', thus motorcycles were not permitted; and it had restricted access, with one exit near Brühl. It was planned and constructed with four lanes and had a width of twelve meters. Two lanes were reserved for each direction and road markings in the middle of the road separated them. From a traffic standpoint, it was a very innovative concept to construct a four-lane Autobahn. The double lanes were justified with the argument that a second lane in each direction was necessary for overtaking, since trucks only traveled at 40 km/h and caused a slow-down of traffic.[214] This argument must be seen within the context of the forecast for a high volume of traffic on this

211 Grossjohann: Der Beginn einer neuen Epoche in Deutschland – Zur Eröffnung der Autobahn Köln–Bonn am 6 August 1932, in: Die Autobahn, issue 8, 1932, pp. 1–5, ADAC Motorwelt Nr. 33, 1932, p. 13, Hans Bretz*:* Das Rheinland baut Strassen, in: ADAC Motorwelt Nr. 33, 1932, p. 27s.

212 Die Kraftwagenstrasse Köln–Bonn, edited by Rheinische Provinzialverwaltung, Düsseldorf, 1932, Stadtmuseum Bonn.

213 Stadtarchiv Bonn, Image no. 4738.

214 Grossjohann (1929) p. 416.

road. The Bonn–Cologne–Düsseldorf country road was considered the busiest road in Germany at the time, with a peak traffic volume of 1,800 vehicles per hour. Specifically for the opening of this expressway, a special police regulation was issued on 2 August 1932, to limit the usual traffic on country roads to automobiles only. This forbade a stopping or parking of vehicles. Horses-drawn carriages and carts that were pulled by cows were prohibited from using the road as were bicycles, and the road was similarly barred for the herding of animals.[215]

Unlike the Bonn–Cologne *autobahn,* the North Italian expressways were planned only with two to three lanes, with both directions using the middle lane for overtaking. This solution worked quite well since the volume of traffic was very low on these roads. In 1928, an average of 1497 vehicles per day could be counted on these routes – a number close to the maximum of 2073 vehicles per day.[216]

The Bonn–Cologne expressway was opened a few months before the National Socialists took over. The Nazi expressway project would go on to make use of many of the concepts from Adenauer's expressway project. A key role was played by the Hafraba manager, Willi Hof, who gathered extensive information from Adenauer's Bonn–Cologne project and then, in April 1933, proposed an expressway project to Hitler for all of Germany.[217]

As with the Bonn–Cologne project, in the Nazi era the expressway construction was in large part financed by funds from the unemployment insurance thereby linking the transport policy to social policy. The issue of expressway toll-financing was also taken up in the NS project – it was deemed permissible for Reich expressways based on §7 in the law to establish the Reichsautobahngesellschaft. In order to emphasize the priority of the NS expressway project, the Cologne–Bonn expressway was designated as only a country road in Nazi era cartography.

---

215 Kölner Stadtanzeiger on 1 August 2007.

216 Uhlfelder (1930).

217 Letter from Willy Hof to Adenauer on 19 May1932, HAStK, Akte Adenauer 902, Nr. 82, Fasz. 1. Akten der Reichskanzlei, Regierung Hitler 1933–1938, published by Konrad Repgen 1983, p. 305, and report from Hof of Hafraba e.V., Hof,on a meeting with the Reich Chancellor on the planning and financing of autobahn construction on 6 April 1933, ibidem, p. 308.

Image 20: Crash Barriers at the Autobahn Bonn–Cologne[218]

It is remarkable that after 1945 Adenauer never addressed the significant role he played as a promoter for the first German expressway. Thus far, research efforts have not revealed this fact. With some justification one can speak of the Bonn–Cologne expressway as Adenauer's expressway. In his time as the mayor of Cologne, Adenauer was keen on launching a number of projects such as founding the University of Cologne, the establishing the fair and constructing the green belt setting in Cologne. These numerous projects burdened the city's finances enormously and rendered it extremely vulnerable during the Great Depression in 1929. A comparable road policy was followed in the Rhine Province where several projects were realized by credit including the Nürburgring race track, the gradual takeover of communal roads in Provincial Administration and, finally, the three big expressway projects.

218 Die Kraftwagenstrasse Köln–Bonn, edited by Rheinische Provinzialverwaltung, Düsseldorf, 1932, p. 55, Stadtmuseum Bonn.

9.5 Summary

This section reveals the strong influence of the external policy factors in the Rhineland that led to innovative traffic concepts. Expressways and commercial vehicle transportation efforts were exploited to support the manufacturing industries. It becomes apparent that on a Prussian and and state level, the politics had not yet recognized the development potential of an automobile-oriented transport economy. Road tolls were considered backwards-looking and brushed off as “medieval” instead of being acknowledged as an instrument for road financing. The reasons why the controversial §13 could not be restated in the parliament, especially since many initiatives for this came from Hafraba, are unknown. In addition to Frielingsdorf’s survey, here Adenauer’s activities on the Rhine Province level can be documented in detail, and his leading role in the implementation of expressway projects can be seen. As large as Adenauer’s influence in the local politics and provincial politics was, on the Berlin level the necessary funds were not available for him to promote the provincial projects.

The Autobahn Mega Project in the Nazi Era

10 Introduction

The topic of *autobahn* construction during the era of National Socialism seems unusual, considering that Germany was barely motorized in 1933. Repeatedly, it has challenged theorists to interpretation. Stommer and Philipp have designated motorways as the "pyramids of the Third Reich" in order to emphasize the character of National Socialism.[219] Mom highlights the visionary character of the project in its anticipation of mass motorization.[220] Furthermore, the project can be interpreted by a formal analysis with the aim of evaluation research, as has been carried out by the author.

All of these approaches can be integrated demonstratively when the *autobahn* project is related to National Socialist regime's politics of "folk products" – an approach that was developed by Wolfgang König and can be encapsulated as follows:[221] Germany had to make up for a backlog with regard to the important indicators of a consumer society, especially compared to the Unites States, with the help of a federally-controlled economy under conditions of scant resources, a shortage of foreign exchange and frozen wages. A "National Socialist consumer society" was to be developed by providing inexpensive infrastructure, consumer goods, and various services through the promotion of "folk" products for the broad masses. This would correspond to the socialist self-concept of the Nazi-state. This approach has been elaborated by Wolfgang König. He analyses the fields:

1. folk broadcast receivers
2. folk television receivers
3. folk accommodations
4. folk refrigerators

---

219 R. Stommer and Claudia Gabriele Philipp (eds.): Reichsautobahnen – Pyramiden des Dritten Reiches, Marburg 1982.

220 Gijs Mom: Roads without Rails – European Highway-Network Building and the Desire for Long-Range Motorized Mobility, in: Technology & Culture, no. 4, vol. 46 (2005), pp. 745–772.

221 Wolfgang König: Volkswagen, Volksempfänger, Volksgemeinschaft – Vom Scheitern der nationalsozialistischen Konsumgesellschaft, Paderborn 2004.

5. folk vehicles
6. Tourism offers by the organization "Kraft durch Freude" (KDF, Strength through Joy).

König describes how different power centers of the Nazi regime picked up these topics in order to expand their influence. Frequently, there was tension between different authorities on how to define folk products or about the needs of a working-class family, and finally, also on inter-agency competencies. Conflicts also arose with regard to allocating the limited resources of human labor and raw materials for the production of folk products – especially in the light of growing re-armament efforts. However, the construction of the a*utobahn* was carried out until 1938 almost unlimited by armaments production. Apart from the a*utobahn*, only the folk broadcast receiver and the tourist packages of the KDF achieved a similar success among the folk products.[222] All the other attempts to advance a folk product never went beyond a pilot scheme or a post-war vision. This visionary Nazi policy approach for a consumer society also applied to the a*utobahn* project which was to provide an infrastructure for mass motorization. But the start of the war in 1939 put an end to any further implementation of these concepts with the exception, surprisingly, of the a*utobahn*. The KDF package tours were stopped and, while the Volkswagen factory was completed, it was misappropriated for war production. Policy execution was superseded by visions that broadened with the course of the war when victory was still assumed as the only possible outcome. The end of the war, it was thought, would launch a period of abundance through the exploitation of Poland's and Russia's resources. This circumstance, it was hoped, would provide the political and economical foundation for a National Socialist-consumer-society.

König has examined how the National Socialist Party tried to stabilize its regime with its efforts to meet the demands of the consumers in a time of scarce resources. The provision of diversions for the citizenry by forcing the development of tourism and motorization, for example, was also a part of this strategy. Among the mass organizations, the German Labor Front (Deutsche Arbeitsfront – DAF) was particularly significant in this context. The DAF had replaced the socialist and liberal labor unions that had been forcibly disbanded and

---

222 Shelley Baranowski: Strength through Joy: Consumerism and Mass Tourism in the Third Reich, New York 2004.

unified industrial and construction workers. The DAF ran an organization called „Kraft durch Freude“ (KDF or „Strength through Joy“) that sponsored wide-ranging programs for travel, sports, education and entertainment.[223] The KDF organization became very well known through the spectacular propaganda-driven cruises that it conducted to Norway, the Mediterranean area and to Madeira. Nevertheless, at its extensive base, the KDF organized trips within Germany as well as short excursions.[224] Also on offer from KDF were package tours at unbeatably low prices to such destinations as Bavaria's Reit im Winkel or the tourist attractions in Heidelberg. Such excellent conditions were achieved by pushing the railroads for discounted train fares. The train fares reflected a discount of up to 75% and were subsidized by the *Reichsbahn*.

The KDF policies created large-scale mass tourism. And while this policy area was complemented by the NS expressway project, there is no evidence that joint planning was involved. Insofar as studies on KDF are available, they suggest that the transportation in connection with package tours and day-trips was carried out on train rather than buses using the expressways.[225] There are connections between the KDF's tourist activities and the NS expressway project, however, when viewed from the aspect of “landscape consumption.” The term „tourist gaze“ which was coined by the British sociologist John Urry, refers to a tourist's desire for mobility, panoramic views and a distance – both in time and in space – from his or her normal routine.[226] Just as the KDF transformed the landscape into a consumer good for tourist consumption, the Nazi expressway project produced views to satisfy the tourist gaze. The expressway's course was adapted to the landscape so that it might transform the journey into a series of unexpected views for the driver. If one were to also take into account the construction of the Volkswagen plant in Wolfsburg in 1938 – which was directed by the DAF

---

223 Wolfgang König: Nazi Visions of Mass Tourism, in: Laurent Tissot (ed.): Development of a Tourist Industry, Neuchatel 2003, pp. 261–268.

224 Timothy W. Mason: Arbeiterklasse und Volksgemeinschaft, Gütersloh 1975.

225 Christine Keitz: Reisen als Leitbild – Die Entstehung des modernen Massentourismus in Deutschland, Munich 1997, pp. 223–232. KDF provided diversions for the expressway laborers. Theater productions were organized in the work camps on a monthly basis, see Kraft Durch Freude, Gau Kurhessen, Program February 1937 through April 1937, library, University of Marburg.

226 Shelley Baranowski: Tourism Industry and Modernity: Comments, in: Laurent Tissot (Ed.): Development of a Tourist Industry, Neuchatel 2003, p. 25. John Urry: The Tourist Gaze: Leisure and Travel in Contemporary Societies, London 1990. The concept of landscape consumption was introduced by Thomas Zeller: Driving Germany, 2007, p. 138.

– then it appears as though a coherent strategy was in place; a three-pronged mechanism comprising the KDF, Volkswagen and the expressways. Research, however, has brought to light that the connection between the Volkswagen plant and KDF was merely the result of a random process and not a joint strategy.[227] Die KDF program of the "Gau Kurhessen" from February to April 1937 gives no indication that the excursion to Heidelberg on the Frankfurt–Heidelberg expressway was at all considered to be a journey highlight.

After the downfall of the Third Reich in 1945, something happened that a*utobahn* constructors had not expected until centuries later: The a*utobahn* project became a symbol of National Socialism. In the postwar period of the 1950s and 1960s, the truncated a*utobahn* network was regarded as a genuine justification for the National Socialist state in the divided German nation.[228]

The German *autobahn* project reflects many aspects of the National Socialist regime's policies, in particular its unbridled use of force both internally and externally. The head of the *autobahn* project, Fritz Todt, was not only a road construction engineer, but also a staunch National Socialist who had joined the party as early as 1923 and who cannot, therefore, be regarded as an apolitical technocrat. On the contrary, Hitler took pleasure in introducing him as an old comrade. It was he who conducted the forced recruitment of the unemployed for *autobahn* construction as a National Socialist political project and also organized the labor camps for construction workers strictly in accordance with the principles of Nazi ideology. Work was compulsory and laborers were unable to escape the heavy physical job which, in the first year, had to be carried out with virtually no help from machinery.[229] For that reason, autobahn labor camps had much in common with concentration camps. The autobahn project closely followed Hitler's expansionary foreign policy. Planning for *autobahn* construction in

---

227 Hans Mommsen and M. Grieger: Das Volkswagenwerk und seine Arbeiter im Dritten Reich, Düsseldorf 1996. Heidrun Edelmann: Der Traum vom "Volkswagen", in: Hans-Liudger Dienel and Helmuth Trischler (Ed.): Geschichte der Zukunft des Verkehrs, Frankfurt 1997, pp. 280–289.

228 „In mancher Hinsicht ist die deutsche Autobahn ein keineswegs geringeres Nationalsymbol und Monument als das Brandenburger Tor selber" (In some ways the autobahn is no less a national symbol and monument than the Brandenburg Gate itself.), James E. Young: Erinnerung ist eine endlos befahrbare Strasse, in: Rudolf Herz and Reinhard Matz: Zwei Entwürfe zum Holocaust-Denkmal, Nürnberg 2001, p. 79.

229 See E. Schütz and E. Gruber, Mythos Reichsautobahn, Berlin, 2nd edition, Berlin 2000.

Austria was started in 1938, immediately after that country's annexation.[230] And the planning for Czechoslovakia and Poland was started in 1939, the year these countries were invaded.[231] Finally, it should be remembered that the *autobahn* project was also one of Hitler's favorite projects and he enjoyed being presented with the details and personally making final decisions about routing and construction.[232]

---

230 See Kreuzer, Bernd: Tempo 130. Kultur- und Planungsgeschichte der Autobahnen in Oberösterreich, Linz 2005.

231 The expansion is documented in "Die Strasse" in the years 1938 till 1942. For the projected route Breslau–Brünn–Vienna through Czechoslovacia see the treaty between Germany and Czechoslovacia on 19 November 1938, in: RG 1938, II, p. 236s.

232 Speer reports on excursions of Hitler to the building sites of the autobahn, in: Richard Overy: Interrogations – The Nazi Elite in Allied Hands, London 2001, p. 229. Windisch-Hojnacki, (cf. appendix 2), documents the selection of the designs in the cases of the Mangfall Bridge and of recreation area Chiemsee by Hitler, p. 97, 147.

11 The Goals of Autobahn Construction within the Context of Transport Policy

The goals of autobahn construction will be examined using the methods of evaluation research. The explicit and the hidden goals are evaluated and the change in goal priorities during the execution of the autobahn project is considered. With this analysis, the question of how the Nazi ruling-system should be perceived is raised. Was it consistently geared towards the dictator Hitler? Were there coherent policies in the various political fields? How did the power centers of the polycratic ruling system articulate their interest in increasing their own power and influence?

The German expressway project was an "ideal" National Socialist project insofar as it bundled several functions:

- Based on HAFRABA plans, it could be started immediately in Summer 1933 with the route Frankfurt a. M.–Darmstadt.
- It combined labor and economic policy targets, as well as tourist, military and transport economic targets.
- It conveyed the power and effectiveness of the new regime and provided a wide scope for propaganda. Schütz and Gruber's study has exposed, in particular, the propaganda function of the expressway. To overcome the humiliation that the German population had suffered through World War I, the Treaty of Versailles, inflation and the Great Depression, "monumental symbols were needed to signify greatness, stability and permanence", namely the well-known massive constructions of the NS regime – the expressway project being, however, the very first.[233]
- Hitler could exploit the utopian character of the Hafraba project and an extensive expressway system in a country with a low car density for his own objectives.

One could say that the autobahn project, with its bundling of so many different interests, became the "ideal" National Socialist project – especially with regard to establishing the

233 Schütz and Gruber, op. cit., p. 13. Schivelbusch points out that monumental constructions were part of the "international style" of the 1930s in: Schivelbusch: Entfernte Verwandtschaft – Faschismus, Nationalsozialismus, New Deal 1933–1939, Munich 2005, p. 13.

regime's reputation at the beginning of its rule. These individual points will be examined below, whereby the author will also try to demonstrate the continuity of the individual goals of the debates of the 1920s. The execution of the autobahn project revealed deep contradictions about rearmament policies, however.

The many interrelated goals that were formulated in the NS expressway project led to its being perceived as ambivalent. The lack of democratic debate in public arenas, which would have resulted in the formation of certain priorities, makes it difficult to assess the importance of the goals. And one can observe some tensions between the *autobahn* project and other interests of the regime; the four-year rearmament plan that was begun in 1936 competed with the expressway project for the limited resources of workforce and steel. Ambivalence can also be seen in the providing of a network of high performance roads but limiting its use for cargo transport, indeed, the transport administration restricted the use of trucks and ran in a transport crisis in 1938.[234]

11.1 The Goal of Political Propaganda

If all objectives were to be weighted, the propaganda aim is clearly the most prominent. Here a link could be made to the visionary projects in the transport sector of the time before 1933: the discussion about the Hafraba expressway route from north to south Germany and the propagandistic exploitation of monumental projects abroad, such as the Großglockner Alpine Road in Austria and the Belomor Canal from Leningrad to the White Sea.[235]

When in 1933 the National Socialists adopted Hafraba's strategy and began detailed planning and construction of the *autobahn* network, they were able to unite two goals: Job creation and propaganda. *Autobahn* construction could be presented as a magnificent measure to create jobs in the spring and summer of 1933 after the number of registered unemployed had climbed to six million in January of that year. The Nazis strongly emphasized the goal of job creation in their propaganda. In his speech on 1 May 1933, Hitler announced, "We will

234 Heidi Rohde: Transportmodernisierung oder Verkehrsbewirtschaftung?, Frankfurt 1999. Vahrenkamp 2008.

235 Rigele, G.: Die Großglockner Hochalpenstraße. Zur Geschichte eines österreichischen Monuments, Vienna 1998. Schlögel, Karl: Sankt Petersburg – Stadt am Meer, Stadt am Belomor-Kanal, in: Karl Schlögel: Marjampole, Munich 2005, pp. 141–154.

continue to endeavour to realize large public job-creation schemes this year," and referred to "our new road construction program, a gigantic task that requires billions."[236] After the act to establish the "*Reichsautobahngesellschaft*" in June 1933, the *autobahn* construction project was announced to the public by the media.[237] Between 25 June 1933 and 22 July 1933 the party newspaper "*Völkischer Beobachter*" featured four front-page reports on the project, including one on 12 July 1933 about the "great *Reichsautobahn* that, at the suggestion of the *Führer* Adolf Hitler, will be created to combat unemployment and promote the automobile industry."[238] The pompous groundbreaking ceremony attended by Hitler on 23 September 1933 to mark the start of construction on the section Frankfurt–Mannheim was massively exploited by the media for propaganda purposes.[239]

In the spring of 1934 (21 March), the expressway construction began at 22 locations simultaneously.[240] At the opening ceremony speech in Bavaria on the same day, Hitler placed great stress on the aspect of job creation.[241] In Baden, the 21st of March was celebrated with large rallies in Mannheim, Heidelberg and Karlsruhe. "With animated music, the workers paraded as a united procession through Neckarstadt (Heidelberg), which was adorned with flags and resplendent. The SA (*Sturmabteilung* – paramilitary branch of the Nazi party), SS (*Schutzstaffel* – paramilitary and security organization), Labor Service, and police force stood in rank and file, as well as the host of unemployed, above whom banners were stretched with the words 'away from unemployment and misery, Hitler will lead us this day'", as the press of Baden reported.[242] In his pubic address, Baden's Minister President Walter Köhler compared –

---

236 Compare Hitler's speech on 1 May 1933, in M. Domarus: Hitler – Reden und Proklamationen, volume I, Wiesbaden 1973, p. 263.

237 "Das neue deutsche Autobahn-Netz – Ein Teil des neuen Arbeitsbeschaffungsprogramms," was the headline on page 3 of the Kasseler Post on 22 June 1933, UBK. "Der gigantische Autostraßen-Plan" was the headline on the front page of the Hessischen Volkswacht, the newspaper of the NS Gau Kurhessen, on 26 June 1933 UBK. The front page of Der Völkische Beobachter on 25 June 1933 (Ausgabe A) reported: "Riesenprojekt der Reichsregierung: Kabinett Hitler beschließt Bau von Reichsautobahnen„. (Deutsche Bücherei, Leipzig).

238 Ausgabe A, Deutsche Bücherei, Leipzig.

239 Schütz and Gruber , cf. appendix 2, pp. 41–43.

240 Die Autobahn, 1934, p. 168, in Spring 1934, on 21 March 1934, Goebbels arranged for the start of constructions to be coordinated throughout the Reich in 22 construction sections, see letter on 12 February 1934 from Thuringia Department of the Propaganda Ministry to the president of Thuringia, in: Thüringisches Hauptstaatsarchiv Weimar, C234, p. 19. Hitler's speech at the construction section Unterhaching near Munich was transmitted by radio to the workers at the other construction sites.

241 „Neue Hoffnung für Millionen" was the headline on the Kasseler Post of 22 March 1934, UBK.

242 Press of Baden on 21 March 1934, Stadtarchiv Karlsruhe, 1/H-1596.

with unintended humor – the expressways to "eternal values" like the Egyptian pyramids and the grandiose constructions of the Romans and Greeks. The following image shows a procession of workers in Karlsruhe on this date.[243]

Image 21: Manifestation in Karlsruhe on 21 March 1934

The propaganda goal of portraying the autobahn project as a job creation program must have had a far-reaching effect within the Third Reich. The files on *autobahn* construction in the Federal Archive in Berlin contain countless letters from *Gauleiter* (leader of a Gau of the NS party) and *Landräte* (administrative district leaders) to Todt and the *Reichsautobahngesellschaft* for the years 1933 and 1934 requesting a rapid start of construction to alleviate the great poverty in these regions.[244]

---

243 Ibidem.

244 For example the letter from the Gauleitung of Kurhessen on 6 July 1935 to the Direction of the Reichsautobahn requesting a rapid start to construction in Knüll, as the population there was extremely needy, in Bundesarchiv R4601, volume 4043.

Regarding the propaganda aspect, Hitler assumed that the planning of these major roads would have a psychological impact on the population and serve to build up trust[245] – an important consideration for safeguarding Nazi rule in 1933. Furthermore, *autobahn* propaganda efforts, addressed such diverse intentions as veiling the backward-looking ideas and aggressive objectives of National Socialism beneath a modern transport concept and, thus, also satisfying those parts of the Nazi movement that were pushing for modernization. For this combination of a backward ideology and modernization, Herf coined the term "reactionary modernism".[246] Todt, the General Inspector for German Roads, who was appointed by Hitler, can be considered a representative of this modernizing segment of the movement. He had already issued a memorandum about the inadequacies of the German road network in December 1932.[247] *Autobahn* construction enabled the propagandists to demonstrate the government's determination to push through new projects – the seemingly endless concrete ribbon was visible to all, and its modernity proclaimed the dawn of a new era.[248]

The goals of job creation and propaganda received significantly greater weight by Hitler's presentation of the *autobahn* concept as a **network**, cf. Image 29.[249] This approach differed from the Hafraba publications which focused largely on the north-south route (Hamburg–Hanover–Kassel–Frankfurt–Basel). Hitler recognized that the propaganda value of an *autobahn* network throughout the Reich was far more effective as a tool to tout job creation and a new age than a mere north-south route from Hamburg to Basel.[250] Hitler also explained his

---

245 In a discussion concerning the financing of the autobahns on 18 September 1933, Hitler referred to "the psychological effect that the construction of these large roads would have. Work on such a massive scale as this would have a confidence-building effect," see: Akten der Reichskanzlei, Regierung Hitler 1933–1938, published by Konrad Repgen, Teil I: 1933/34, volume 2, edited by Karl-Heinz Minuth, p. 742, Boppard am Rhein 1983.

246 Jeffrey Herf: Reactionary Modernism. Technology, Culture, and Politics in Weimar and the Third Reich, Cambridge, 1984, Chapter 8.

247 Todt's study was titled "Straßenbau und Straßenverwaltung", BA, R 4601, volume 3001/A.

248 On this interpretation see also Shand 1984, cf. appendix 2.

249 The picking up of the network idea is documented in an annotation by Ministerialrat Willuhn concerning a discussion between the Chancellor and the director of the HAFRABA e.V., Hof, about the planning and financing of autobahn construction on 6 April 1933, in: Akten der Reichskanzlei, Regierung Hitler 1933–1938, published by Konrad Repgen, Teil I: 1933/34, volume 1, edited by Karl-Heinz Minuth, p. 305, Boppard am Rhein 1983, and a report of the director of the HAFRABA e.V., Hof, about his discussion with the Chancellor about the planning and financing of autobahn construction on 6 April 1933, op. cit., p. 308. Also on the network idea, the Chancellor's discussion with leading industrialists on 29 May 1933, op. cit., p. 511.

250 For propaganda reasons on 21 March 1934 the autobahn project was started at 22 sections simultaneously, Die Autobahn, 1934, p. 168. This start was conducted by Goebbels, see the letter on 12 February 1934 from the Thüringen section of the propaganda administration in Berlin to the President of the state Thüringen, in: Thüringisches Hauptstaatsarchiv Weimar, C234, p. 19.

preference for the network on the grounds that it would be easier to push through a large project than individual road sections against the opposition of special interests groups.[251] The network concept also had the advantage of allowing construction to start simultaneously at several different locations within the Reich, and thus, providing a far greater number of propaganda opportunities than, for example, constructing two large hydroelectric power plants in order to create jobs.[252]

According to Schütz and Gruber, the *autobahn* project gave the National Socialists an opportunity to orchestrate the presentation of the Third Reich.[253] Thus, 90 000 members of Nazi mass organizations were commandeered to line the streets for Hitler's triumphal procession from Frankfurt to Darmstadt when he inaugurated the first section of the *autobahn* on 19 May 1935. Hitler arrived in Frankfurt by air accompanied by Goebbels, Schacht and the head of the press office, Dietrich, and gave Todt a Mercedes cross-country vehicle as a sign of his appreciation.[254] One year later, the inaugurations of other sections were regularly co-ordinated to coincide with special occasions. On 17 August 1936, five sections were scheduled to open simultaneously at the end of the Olympic Games in Berlin to ensure the greatest possible publicity from the international press corps that had gathered in Berlin to cover the Games.[255] On Sunday, 27 September 1936, the third anniversary of Hitler's first ground-breaking, 17 sections were opened to complete the first 1000 kilometers.[256] *Autobahn* construction was also incorporated into the programs of National Socialist Party conventions.

---

251 See the discussion with the director of the HAFRABA, Hof, on 6 April 1933 and with the director general of the Deutsche Reichsbahngesellschaft, Dorpmüller, on 10 April 1933, in: Akten der Reichskanzlei, Regierung Hitler 1933–1938, op. cit., p. 305, and p. 331.

252 Kaftan, K.: Der Kampf um die Autobahnen, Berlin 1955, makes this comparison, p. 154.

253 Schütz and Gruber, op. cit.

254 See archive of the city Kassel, Kasseler Neueste Nachrichten on 20. 5. 1935, p. 1.

255 Statistisches Jahrbuch für das Deutsche Reich, 1936, p. 198. Todt's press statement in the Kurhessischen Landeszeitung on 18 August 1936, UBK, in which he referred to the traffic induced by the Olympic Games. The issue 14, the 2nd issue in July 1936, of the journal “Die Strasse” focuses on the Olympic Games.

256 Statistisches Jahrbuch für das Deutsche Reich, 1936, p. 198. "Deutschland dankt jubelnd dem Führer – 1000km Autobahn fertig" was the headline on the front page of the Kasseler Post on 28 September 1936 UBK.

In 1937 and 1938, from the speaker's podium at the NS party's main congress in Nuremberg, Todt delivered progress reports on the construction achievements thus far.[257]

11.2 The Goal to create Jobs

In his speech before the construction workers in Unterhaching on 21 March 1934, Hitler emphasized the aspect of job creation in connection with the expressway project.[258] Providing jobs for the unemployed was not an original idea of the National Socialists – this concept had existed since the 1920s. Investments in the transport sector for the creation of jobs – whether canal building, *Reichsbahn* routes, or road and expressway constructions – had already been the subject of many discussions in the cabinets of Brüning and Papen.[259] In the Brüning cabinet the construction of the expressway Frankfurt–Mannheim was already being prepared. Job provisioning was also a main argument for road construction projects on state and province levels.[260] The Cologne–Bonn stretch was carried out from 1929 to 1932 with minimal use of machinery to leave as much work as possible for the unemployed and was also partly financed through unemployment insurance funds (cf. section 9). Expressway construction for jobs was an idea that was also propagated during the Great Depression in both of the international expressway conventions of Geneva and Milan in the year 1931 and 1932.[261] In this respect, the expressway concept benefited from the world economic crisis. There was also

---

257 F. Todt: The General Inspector for German Roads presented a report at the Nazi party convention in Nuremberg 1937 about "Bau und Verkehr auf den Strassen Adolf Hitlers", in: Die Strasse, issue 18, 1937, pp 518–520. and F. Todt: Der Generalinspekteur für das deutsche Strassenwesen berichtet auf dem Parteitag Großdeutschlands über die Strassen des Führers als Schrittmacher der deutschen Bauwirtschaft, in: Die Strasse, issue 19, 1938, pp.600–603. In both speeches he legitimizes autobahn construction with a reference to traffic over the Whit holiday.

258 M. Domarus: Hitler – Reden und Proklamationen, Volume I, Wiesbaden 1973, p.371.

259 See e.g. the minister meeting of 19 May 1930 in: Akten der Reichskanzlei Weimarer Republik, published by Karl Dietrich Erdmann, Kabinette Brüning I a. II, volume 1, 30 March 1930–10 October 1931, edited by Tilmann Koops, Boppard am Rhein, 1990, pp. 137–140, or also the cabinet meeting of 5 June 1930, where the construction of the Hafraba route was discussed, in the place cited, pp. 188–192. See also H. Marcon: Arbeitsbeschaffungspolitik der Regierungen, Papen and Schleicher, Frankfurt 1974.

260 So e.g. Upper Hesse and Saxony in the year Jahre 1930, see ADAC-Motorwelt, 1930, nr. 43/44, p. 37.

261 Schipper 2008, pp. 83–120. Hafraba-Mitteilungsblatt, issue 9, 1931, pp. 1–2, Die Autobahn, issue 4, 1932, p. 7, and issue 5, 1932, pp. 1–2. Because according to the plans of the Hafraba, the expressway project was to be financed through toll fees, the profitability of the project played a major role. Employing the jobless to construct the expressway gained a new focus when it was brought forward that this would sink construction costs, see the report to the 2. International Expressway Congress in Mailand, in: Mitteilungsblatt, issue 5, 1932, p. 1. See also the section „Die arbeitswirtschaftliche Bedeutung für den Arbeitsdienst" (the economic importance of work service) of the Hafraba application from August 28th 1932 to the Reichsverkehrsministerium for the granting of a special permission after §13 of the Law on Revenue Sharing for the construction of the expressway Frankfurt–Mannheim, *Kaftan*, in the place cited, p. 147s.

the Gerecke plan for job creation which, among other things, demanded the construction of "fast transport connections of all kinds"; it was publicly debated and covered in the December 1932 issue of the HAFRABA magazine.[262]

Concerning the claim of the propagandists that the expressway project noticeably contributed to reducing the rate of unemployment, however, there is a considerable discrepancy. Schütz und Gruber were able to prove that the overall scope of employment created by the expressway network was in fact quite limited:[263] In August of 1934, approximately 50 000 workers were directly involved in the construction work. This number climbed to 112 000 in May 1935 and its highest level of 125 000 was reached in July 1936. In the years between 1937 and 1939, the numbers leveled off to about 80 000–110 000. Suppliers, in turn, also employed roughly the same number of workers. The minimal effect that the expressway project – with its 50 000 laborers – had on the labor market in the summer of 1934, is revealed by the fact that the number of unemployed had already dropped from six million in January 1933 to 2.5 million by the summer of 1934.

11.3 The Goal to promote Motorization

The National Socialists also had another economic policy goal connected with the *autobahn*, namely to increase the degree of motorization in Germany which was clearly underdeveloped by international standards.[264] In chancellery discussions, Hitler explicitly linked *autobahn* construction with the promotion of the automobile industry that he accurately predicted would be "the most formidable and successful of the future."[265] Undoubtedly the existence of high-

262 Die Autobahn, issue 12, 1932, pp. 1–3. Gerecke, president of the "Deutschen Landgemeindetages", had transmitted "the guidelines for a job creation program to remove unemployment" to the Reich Chancellor von Papen in August 1932 and was invited to a talk with the chancellor on September 5th 1932, see records of the Reich Chancery, Das Kabinett von Papen, volume 1, edited by Karl-Heinz Minuth, Boppard on the Rhine 1989, p. 513. In December 1932, Gerecke was appointed Reichskommissar for job creation.

263 Schütz and Gruber, op. cit., pp. 56–58.

264 Vehicle density per 100 inhabitants was only one quarter of that of France or England, see Statistisches Jahrbuch für das Deutsche Reich 1933, p. 84*. The promotion of the automobile industry is stated in the grounds for the law of 27 June 1933 for the establishment of a Reichsautobahn Company (Reichsautobahn-gesellschaft), in: Die Autobahn, issue 7, 1933, p. 2.

265 See his meeting with leading industrialists on 29 May 1933, Akten der Reichskanzlei, Regierung Hitler 1933–1938, op. cit., p. 511, p. 513. He also argued that multiplyer effects would stimulate the economy, a very modern sounding argument for this period. Interesting because the fixation upon war from the very beginning is documented – during this meeting Hitler already anticipated air attacks against the autobahn and recommended reinforcing the concrete, op. cit., p. 511.

speed roads provides a strong incentive – in the medium term – for private households and businesses to purchase vehicles by promising significantly reduced traveling times and greater flexibility compared to rail transport.[266]

The promotion of the automobile industry should also be seen in terms of the demand for cars in early 1933. On 1 July 1932, there were 497 000 cars and 157 000 trucks licensed in the Third Reich – a low level of motorization, particularly compared to the USA, where there were 25 million vehicles. Car production in Germany fell from 108 000 in 1928 to a low of 43, 000 in 1932 during the Great Depression, before rising again to 92 000 and then 147 000 in 1933 and 1934. According to Henning's[267] analysis, from October of 1932, the number of registrations had already started to rise sharply. He ascribes this growth to the trust in the government's economic policies, which began to manifest itself at the conclusion of the Conference of Lausanne in June 1932, and which was reinforced by the Papen Economic Program that was announced at the end of August 1932. The increase of car registrations was stimulated by a tax exemption for new car purchases from March 1933. Beginning in March 1933, the car industry and its suppliers reported high capacity utilization and new hiring. In his meetings in the Reich Chancellery, Hitler went beyond the current situation in the car industry and formulated a long-term target of three to five million cars which he hoped to achieve through expressway construction.[268]

The desire to stimulate demand for cars can be seen in the context of the transport policy debates between 1930 and 1932. An analysis of the journals on transport policy between 1930 and 1932, shows that repeated reference was made to the economic importance of the

---

[266] Woerner estimates that it took 1.5 hours to travel from Munich to the famous ski region Reit im Winkl by car on the autobahn, compared to 5 hours by rail, see Anton Woerner: Die Reichsautobahn Munich–Salzburg, in: Die Autobahn, issue 9, 1933, pp. 10–12.

[267] H. Henning: Kraftfahrzeugindustrie und Autobahnbau in der Wirtschaftspolitik der Nationalsozialisten 1933 bis 1936, in: Vierteljahresschrift für Sozial- und Wirtschaftsgeschichte, vol. 65, 1978, pp. 217–242, see also P. Krüger: to Hitlers' ‚national socialist knowledge of economy', in: Geschichte und Gesellschaft, vol. 6, 1980, pp. 263–282.

[268] See meeting at the Reich Chancery with transport experts on 18 May 1933, Repgen, Konrad (ed.): Akten der Reichskanzlei, Regierung Hitler 1933–1938, Teil I: 1933/34, Band 1, p. 464 and see his meeting with leading industrialists on 29 May 1933, op. cit., p. 511, p. 513.

automobile industry.[269] This was particularly true in connection with the experiences of the much more highly developed automobile industry in the United States,[270] where the materials and components industries accounted for a large percentage of those employed in the automobile industry. This argument was applied to Germany and demands were made that more be done to stimulate the automobile industry. When Henning[271] asks why Hitler, who was an economic layman, used promotion of the automobile industry as a means of stimulating the economy, attention should be drawn to this transport policy debate. The tax exemption announced in April 1933 for the purchase of new vehicles – to take effect retroactively from March 1933 – should also be viewed against the background of this debate. In the transport policy press of the years 1930 to 1932, many articles were published, especially in the weekly *ADAC-Motorwelt*, which deplored the high tax burden on car drivers and conjured the throttling of the car industry that was contributing to the slow-down in growth. The magazine *Auto und Wirtschaft* reported that tax revenue within the Reich had risen from 275 million RM in the year 1929 to 466 million RM in 1931.[272]

The policy of motorization in the NS-state must also to be evaluated in the context of the discussion about mass production and rationalization in Germany in the 1920s.[273] The German automobile industry was underdeveloped in production engineering compared to the USA whereupon a discussion was sparked off about the need of a folk vehicle. Yet, there was a strong movement favoring the implementation of assembly line production. The concept of the Fordism, which connected mass production and mass wealth, was widely known.[274] This is illustrated by the large number of editions of Henry Ford's autobiography "My life and

---

269 See also for instance August Müller: Das Auto als Konjunkturhebel, General-Anzeiger, Frankfurt a.M., 11 July 1930, quoted from the HAFRABA-Mitteilungsblatt, 1930, issue 8, p. 19. See also; Reichsverband der Automobilindustrie (Editor): Tatsachen und Zahlen aus der Kraftfahrzeugindustrie, Berlin 1931.

270 ADAC-Motorwelt had 14 reports about the USA in 1930, e.g. about the prevalence of cars in the USA in issue 1, p. 22–24. Die Automobilwirtschaft in den USA, in: Wirtschaftsdienst, 16. Jahrgang 1931, pp. 1079–1081.

271 Henning, see appendix 2, p. 221.

272 Dr. H. Kruse: Die Notlage der Kraftverkehrswirtschaft,in:Auto und Wirtschaft, nr. 1,Berlin 1932, p.1.

273 Reiner Flik: Von Ford lernen? Automobilbau und Motorisierung in Deutschland bis 1933, Köln 2001, also: Richard Vahrenkamp: Von Taylor zu Toyota – Rationalisierungsdebatten im 20. Jahrhundert, Cologne 2010.

274 Kurt Becker, Frank: Autobahnen, Auto-Mobilität. Die USA; Italien und Deutschland im Vergleich, in: Politische Kulturgeschichte der Zwischenkriegszeit 1918–1939, Göttingen 2005 (Geschichte und Gesellschaft Sonderheft 21), pp. 23–59.

Work" which was published in German from 1923; the 33rd edition appeared in 1940. Hitler adopted the concepts of industrial mass production and folk motorization from Ford and founded the Volkswagen factory in 1938 despite massive opposition from the German automobile industry.[275] In the same year, he decorated Ford with the medal „*Großkreuz des Deutschen Adlerordens*" (or Grand Cross of the Order of the German Eagle). The ideology of anti-Semitism also connected Hitler and Ford. Hitler also drew on Ford's pamphlet on supposed "the danger of international Judaism", which was released as a German edition in 1921 in Leipzig for his book "Mein Kampf".[276] Hitler had a similarly close personal relationship with another automobile executive – Jakob Werlin of Daimler-Benz, who had been appointed by Hitler to handle automobile policies. He joined Hitler in partaking in a ceremony on the occasion of the first 1000 autobahn-kilometers at Breslau on 27 September 1936. Hitler mentioned Werlin explicitly, in addition to Porsche, as an important promoter of the Volkswagen-project in his speech to commemorate the laying of the foundation stone of the Volkswagen factory on May 26th, 1938.[277]

The promotion of motorization during the Nazi era up until 1939, was in constant conflict with re-armament efforts over the scanty resources. However, promotion of motorization was not only an unsubstantial statement but was realized with different instruments. The main instruments which contributed to a vehicle-friendly atmosphere, were the 1933 tax exemption on car purchases, the *autobahn* construction, the touring exhibition "The Street" ("*Die Strasse*"), and the activities of the National Socialist automobile corps.[278] In many cases, big events were staged as motor-sport happenings such as the 2000 kilometer-long journey in July 1933 and the car rally to Tannenberg and Königsberg at the end of August 1933. Goebbels

---

275 Hans Mommsen and Manfred Grieger: Das Volkswagenwerk und seine Arbeiter im Dritten Reich, Düsseldorf 1996, and: Wolfgang König: Volkswagen, Volksempfänger und Volksgemeinschaft, Paderborn 2004.

276 Baldwin, Neil: Henry Ford and the Jews, New York 2001, Chapter 12. Mommsen and Grieger: Das Volkswagenwerk und seine Arbeiter im Dritten Reich, Düsseldorf 1996. In the 33th edition of the autobiography of Henry Ford there is a photo of Ford decorated with the Grand Cross of the Order of the German Eagle.. The role of the American and German Ford plants in the motorization of the German Army is described in the official report (without an author) of the Ford Motor Company „Research Findings about Ford-Werke under the Nazi Regime", p. 26–28, 2001 in: www.Media.Ford.com.

277 Hans Pohl, Stephanie Habeth and Beate Brüninghaus: Die Daimler-Benz AG in den Jahren 1933 bis 1945. Eine Dokumentation, Zeitschrift für Unternehmensgeschichte, Beiheft 47, Stuttgart 1986, and: Kasseler Post on 28. September 1936, UBK. Völkischer Beobachter on 27 May 1938, in: Pohl u.a., op. cit., document 21.

278 Dorothee Hochstetter, Motorisierung und Volksgemeinschaft, Munich 2005.

had promoted it as the "Eastern journey of loyalty" ("*Ostland Treuefahrt*") on the occasion of the 19th anniversary of the battle of Tannenberg. 1045 automobiles and 483 motorbikes from National Socialist groups and automobile clubs from throughout Germany took part (cf. section 20).[279] The promotion of automobile racing was reviewed by Reuss.[280]

The extensive criticism of the Volkswagen project, included charges, that it had only raised unfulfilled hopes for a grand future with automobiles, fails to appreciate the dimension of motorization promotion. The intensity of motorization policy becomes apparent by showcasing the development of the vehicle population of Frankfurt on the Main.[281]

Year	Automobiles and buses	Automobiles and buses per 10,000 inhabitants	Motorcycles	Motor vehicles overall (including trucks)	One motor verhicle per ... inhabitants
1927	4,278	90.2	3,655	9,484	50
1932	7,912	148.0	6,411	16,705	32
1938	18,542	334.4	11,226	34,658	16
Increase 1932–1938 (percent)	134.4		75.1	107.5	
Increase 1927–1932 (percent)	84.9		75.4	76.1	

Table 5: Motorization in Frankfurt on the Main from 1927 until 1938

The data in Table 5 shows the increase in the vehicle population and registered motorcycles. In both periods, 1927 to 1932 and 1932 to 1938, which can be compared, the increase of the vehicle population was clearly greater than the increase for registered motorcycles. In the second period, the process actually accelerated from 84.9% to 134.4% concerning the vehicle population. This indicates that the motorization policy was successful even if automobile density did not reach the levels seen in France or Britain. A comparison of the growth rate for automobiles and motorcycles shows that there was a strong desire to own an automobile. Likewise, the growing number of driving licenses emphasized the successful implementation

279 Frankfurter Nachrichten on 23 July 1933 and 26 August 1933, Institut für Stadtgeschichte, Frankfurt. Zur Ostland Treuefahrt see section 20.

280 Eberhard Reuß: Hitlers Rennschlachten, Berlin 2006.

281 Statistische Jahrbücher für das Deutsche Reich, 1928 to 1938.

of the motorization policy. Within the license class 3, the number of tests taken rose from 339 thousand in 1935 to 377 thousand in 1938.[282]

One aspect of motorization is the trend of transporting goods by truck as an alternative to rail freight transport. This had been a avowed goal of Todt who had emphasized that 38 000 communities did not even have a railroad connection. However, Todt´s truck transport plans could not be realized because of resistance from Dorpmueller, President of the *Reichbahn* – see below section 11.6.

The National Socialist leader of farmers of the Third Reich, Darré, fixated his ideology on "blood and soil" agricultural policy[283] but failed to promote the sale of tractors and agricultural machinery which would have achieved two goals:

- higher employment in the industry of agricultural machinery.
- higher production of agricultural products, which were urgently needed due to undersupply.

11.4 The Goal to promote Tourism

Apart from the "hard" goals, the propagandists also introduced a whole series of "soft" goals regarding expressway construction:

- The project would make a major contribution for the future of transport.
- It would build a monument to National Socialism that would stand for centuries.
- "Adolf Hitler's roads" would be built.
- Principles of landscape aesthetics would be realized. The *autobahn* would be adapted to the landscape. At the same time, the landscape was to be cared for.
- An ideology was adopted in order to reconcile technology and nature.

We will have a look at how tourism and recreation were part of the *autobahn* propaganda efforts and how the population accepted the new opportunities for recreation. The beginning

---

282 Statistisches Jahrbuch für das Deutsche Reich, 1937, p. 203 and 1939/40 , p. 237.

283 See Grundmann, F.: Agrapolitik im Dritten Reich, Hamburg 1979, G. Fahle: Nazis und Bauern, Cologne 1986, G. Corni: Hitler and the Peasants, New York 1990.

of the *autobahn* project in 1933 already saw proposals on how the autobahn could stimulate tourism in regions where it was poorly developed.[284] These can be traced back to publications in the 1920s when the importance of tourism as an economic factor was recognized.[285]

For propaganda purposes, the peculiar idea of "*Autowandern*" (road touring) was introduced. *Autowandern* was supposed to combine the enjoyment of the landscape with driving through the landscape on the *autobahn*. In his journal "*Die Strasse*", the head of the German *autobahn* project, Todt, published numerous articles to promote *Autowandern*,[286] which was connected with the recommendation to make a picnic on the lawn strip besides the *autobahn* – an idea that has been popular in the Netherlands in the 1960s. Image 22 shows a picnic scene as promoted by Todt's *autobahn* propaganda.[287] However, all these ideas were primarily propagandist because virtually no cars were driven on the *autobahn*, due to the low car density in Germany.[288] Only at holidays, as Easter or Whit Holiday, some excursion traffic could be observed as traffic counts in the years 1937, 1938 and 1939 revealed (see section 24).

In 1936, the propagandists continued their focus on *autowandern* with special issues of "*Die Strasse*", e.g. issue six. The related topic – excursion traffic on weekends – also gained in importance for the propaganda ministry. Already in 1933, the Brandenburg section of Nazi strorm troopers counted the traffic on the arterial roads of Berlin from 23–30 August. They came to the conclusion that traffic doubled on Sunday compared with the average on the six working days.[289] The number of cars and bicycles doubled and motorcycles numbers even tripled depending on regional and social factors. The strongest bicycle traffic originated from the worker's districts on the road to Grünau in a south-eastern traveling direction. Traffic in the western direction to Wannsee was dominated by cars from bourgeois districts. Although this data originated in 1933, it was not published in Todt's journal before 1936. It is

---

284 Anton Woerner: Die Reichsautobahn München–Salzburg, in: Die Autobahn, issue 9, 1933, pp.10–12, and Schmidt-Lamberg, Herbert: Hebung des Fremdenverkehrs durch Autobahnen, in: Die Autobahn, issue 6, 1933, p. 2.

285 Keitz, op. cit.

286 On "Autowandern" see Die Strasse, issues 3, 6, 11, 14, 1936.

287 Front page of issue 6, 1936, of Die Strasse.

288 In 1932 the car density per 1000 inhabitants only amounted to one quarter of that of France and one third of that of England, see Statistisches Jahrbuch für das Deutsche Reich 1933, p. 84*.

289 Census data on the arterial roads in Berlin, in: Die Strasse, issue 6, 1936, p. 178.

interesting that from 1928/29 until 1936 no systematic traffic counts were conducted in the entire Reich. Only the Hansa section of the Gezuvor conducted a traffic survey in 1934 in northern Germany.[290]

Image 22: Picnic at the autobahn

[290] Gesellschaft zur Vorbereitung der Reichautobahnen (ed.): Die Planungsarbeiten für die Reichsautobahnen – Zweieinhalb Jahre GEZUVOR, Berlin 1937, p. 46s. For the traffic count in 1936 see Hoffmann, Rudolf: Die Entwicklung des Kraftverkehrs auf Reichsautobahnen und Reichsstrassen, in: Forschungsgesellschaft für das Strassenwesen (ed.): Strassenbau-Jahrbuch 1937–1938, Berlin, Verlag Volk und Reich, 1938, pp. 29–42.

Another focus of propaganda – the **anticipated** excursion traffic – has been dealt with in Todt's journal. In 1936, an article showed the growth of the regions that inhabitants of the four cities of Berlin, Frankfurt a.M., Munich and Leipzig could reach for a weekend excursion in the years 1880, 1912 and 1938 (when the autobahn network would be rather complete). Associated regional maps were presented for these three years to support a comparison.[291] The aim was to demonstrate the development of wide regions that could be visited for leisure purposes via the *autobahn*. A second article described how the inhabitants of the capital Berlin could reach the popular seaside resorts on the coast of the Baltic Sea and how the completion of the *autobahn* route to Stettin, in 1937, would shorten the journey. Image 23 shows a map of these resorts.[292]

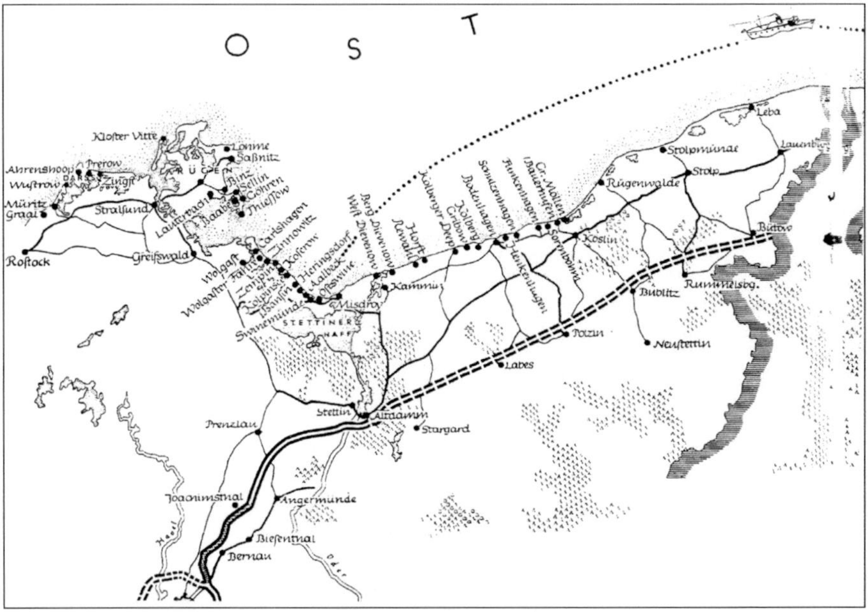

Image 23: Seaside resorts at the Baltic Sea, accessible in the year 1937 from Berlin via the autobahn

291 Ausflugsverkehr der Großstädte, in: Die Strasse, issue 6, 1936, pp.163–167.

292 Von Berlin in die Ostseebäder, in: Die Strasse, issue 6, 1936, pp. 176–178.

11.5 Goals of Military Policy

The military value of expressways for Hitler's warfare has often been questioned.[293] In a speech before the *Reichsverteidigungsrat* (Reich Defense Council) on 18 November 1938, Göring did attribute to the expressways the potential to move troops at a short notice.[294] In fact, however, troop and materials movements were carried out primarily by railroad. The number of cases in which troop movement on the expressways is recorded is small. In 1938 tanks rolled over the expressway while invading Austria.[295] And in 1940, after the invasion of France, tank movements on the Asbachtal Bridge near Bad Hersfeld are recorded.[296]

Ironically, the autobahn served other purposes in military conflicts. In 1945 it facilitated the advance of western allied troops from the Rhineland to the river Elbe[297] and served as the route for civilians fleeing Berlin before the advancing Soviet troops.[298] Finally, the autobahn Hannover–Berlin, that had been in use since 1939, was vital for the ability of Western troops to hold their position in Berlin during the Cold War.

The general promotion of motorization was also linked to the military policy goal of being able to equip the army with an adequate number of cars and trucks in the event of war. The goal of military motorization stands in the context of the policy of motorizing the armed forces that prevailed throughout Europe between 1920 and 1940. As the degree of military motorization was still low, the military saw private motorization in terms of requisitioning.[299]

---

293 Kopper 2002, p. 32. The use of expressways for troop movement was subject of controversy among military leaders Henning, in the place cited, p. 233, Karl Lärmer: Autobahnbau in Deutschland 1933 bis 1945, Berlin 1975, p. 114, p. 138 , Shand, in the place cite., Ludwig, in the place cited, p. 313s.

294 Timothy W. Mason: Arbeiterklasse und Volksgemeinschaft, Gütersloh 1975, Dokument 152, p. 911.

295 See Guderian: Mit der Panzerwaffe auf den Straßen des Sieges, in: Die Straße, issue 23/24, 1940, p. 594: Guderian stresses the high average speed which could be reached on the expressways "while fully sparing man and machine": „Welche Freude, innerhalb des alten Reichsgebietes zu marschieren!“("What joy to march within the ancient territory of the Reich!").

296 Susanne Hohlmann: Pfaffenwald – Sterbe- und Geburtenlager 1942–1945, Nationalsozialismus in Nordhessen vol. 2, Kassel 1983, p. 37.

297 See for example the diary entry on 29 March 1945 of the US Division Artillery, 6th Armored Division on the Internet at http://members.aol.com/super6th/divarty/SignificantDates.htm.

298 See for example, Erich Kästners escape out of Berlin, in: E. Kästner: Notabene 45, Zurich 1961.

299 Henning, op. cit., p. 232.

And in fact, to prepare the German Army for a campaign against France, 16 thousand trucks were confiscated from the German economy.[300]

Numerous statements by Hitler in the Chancellery point to the goal of requisitioning.[301] At a meeting with leading industrialists on 29 May 1933, Hitler linked *autobahn* construction with the vision of a "Volkswagen," or "peoples' car" that would compensate for the disadvantages of a heterogeneous vehicle fleet for mobilization. In the event of mobilization, soldiers would be able to travel on the *autobahn* to their operational areas in a kind of standardized "Volkswagen." In order to implement this idea, the *Reichsbahn* was to manage the entire transport system as a centralized transport company. It was supposed to order large series of the "Volkswagen" from industry, own the *autobahns* and by means of toll fees regulate competition between the road and rail sectors in a manner that would allow the transition from rail to road traffic, (that Hitler regarded as inevitable), to take place in a controlled form that would not impair the efficiency of the *Reichsbahn* as a company.[302] The idea developed here of a centralized large transport company, was first raised in the debate about road-rail competition and was repeatedly brought up by the *Reichsbahn* after 1930.

To discuss the conflict between railway and truck entrepreneurs regarding truck transport, Hitler summoned the Director General of the *Deutsche Reichsbahngesellschaft*, Julius Dorpmüller, to the Chancellery on 16 March 1933 for a lecture about competition between the *Reichsbahn* and trucking companies.[303] Dorpmüller put forward the arguments contained in the *Reichsbahn* memorandum:[304] It made economic sense to use revenue from the transport of

---

300 Hans-Adolf Jacobsen: Der Fall Gelb, Wiesbaden 1957, p. 195.

301 On requisitioning see the discussion between the Chancellor and transport experts on the question of autobahn construction on 18 May 1933, Akten der Reichskanzlei, Regierung Hitler 1933–1938, op. cit., p. 464, and: Memorandum of Oberregierungsrat Willuhn about a lecture by the Director General of the Deutsche Reichsbahngesellschaft on the issue of competition between the Reichsbahn and road freight traffic on 16 March 1933, op. cit., p. 229.

302 See the Chancellor's meeting with leading industrialists on 29 May 1933, Akten der Reichskanzlei, Regierung Hitler 1933–1938, op. cit., p. 511s. However, the military leadership was divided about the suitability of the autobahn for moving troops, see Henning, op. cit., p. 233, Lärmer, see appendix 2, p. 114, p. 138 and Shand, see appendix 2.

303 Annotation by Oberregierungsrat Willuhn about a lecture given by the General Director of the Deutsche Reichsbahngesellschaft on the question of competition between the Reichsbahn and motorised traffic on 16 March 1933, Akten der Reichskanzlei, Regierung Hitler 1933–1938, cit, pp. 225–231.

304 Reichsbahn und Kraftwagenverkehr, published by the Deutsche Reichsbahngesellschaft, Berlin 1930.

high quality goods to subsidize the transport of bulk goods by rail,[305] and thus eliminate price competition from independent trucking companies that mostly transported high quality goods. Moreover, he argued, the provisions of the Rail Transport Ordinance compelled the *Reichsbahn* to transport all goods and treat them equally, which put the company at a disadvantage with independent trucking companies who could choose the lucrative contracts. Hitler opposed Dorpmüller insofar as he said that the *Reichsbahn* would have to accept that the transport of high quality goods was increasingly being switched to heavy trucks as these permitted door-to-door transport. Promotion of truck traffic was also desirable in order to ensure that Germany had a sufficient number of vehicles in the event of their being needed for national defence and to permit technical progress in engine construction. The *Reichsbahn* could not therefore be allowed to let road transport wither. Hitler initially demanded that the *Reichsbahn* undertake freight transport with trucks as a monopoly itself,[306] thus taking up the idea of a large centralised transport company of the kind the *Reichsbahn* had proposed in its memorandum of 1930. However, he focused more on *Reichsbahn* operations with its own trucks that would replace unprofitable railway lines.

Considering that Hitler was actually an advocate of competition, one might wonder why he was so concerned with protecting the *Reichsbahn* from competition with the trucking industry. One indication that he wished to keep a technical and financially well equipped *Reichsbahn* as a logistical instrument for the war effort is the speech he made on 29 May 1933 to leading industrialists, where, in the context of a discussion about military mobilization, he explained that if the *Reichsbahn* were not protected, it would "be useless for other tasks that cannot be carried out by motor vehicles."[307] These statements can be interpreted to mean that he believed a dense network of *autobahns* combined with mass motorization would only be available as a substitute capacity for rail transport after the point in time at which he planned to start the war. Replacing railway transport, that was based on the domestic fuel source coal, with motor vehicles was, from a military perspective, precarious in that the

---

305 Op cit, p. 64.

306 Annotation by Oberregierungsrat Willuhn about a lecture by the General Director of the Deutsche Reichsbahngesellschaft on the question of competition between the Reichsbahn and motorized traffic on 16 March 1933, Akten der Reichskanzlei, Regierung Hitler 1933–1938, published by Konrad Repgen, Teil I, op cit, p. 229.

307 Akten der Reichskanzlei, Regierung Hitler 1933–1938, op cit, p. 511.

necessary materials rubber and gasoline were extremely scarce in Germany and in Hitler's plans had to be reserved for military offensives.[308] These considerations could also indicate that Hitler wished the *Reichsbahn* to undertake tasks that he did not believe could be performed by motor vehicles.

If one assumes that the important role of the *Reichsbahn* for the logistics in the planned war had been recognized, then one had to notice on the other hand, that investment for maintenance and improvement in the railway did not take place until 1939, as Kopper had shown in his study.[309] In the sector of transportation nearly all the investment had been made in the *autobahn* and road network. So the war had been started with an insufficiently prepared railroad. In this point one can recognize a bad fit of policies in the NS political system.

11.6 The Lack of the Goal Truck Transport

However, the law on long distance freight transport enacted in 1935 did not create a *Reichsbahn* monopoly because Hitler feared that competition would later be eliminated completely.[310] Instead, independent trucking companies had to apply for compulsory membership of the "*Reichs-Kraftwagen-Betriebsververband*" (Council of Truck Operators), which in consultation with the *Reichsbahn* laid down tariffs for transport which were then approved by the Ministry of Transport. The intention behind this supervision by the Ministry of Transport was to prevent the tariffs from being undercut. Long distance trucking transport companies were relieved of the elementary entrepreneurial functions of issuing invoices and handling payments, which were instead carried out by the *Reichs-Kraftwagen-Betriebsverband*.[311] This made it significantly harder to evade the set freight rate. Although Dorpmüller could actually

---

308 Henning, op cit.

309 Kopper, 1995, pp. 408–411. During the 1930s one can observe in the US a modernization of locomotive technologies, see Churelle 1998.

310 F. Hartmansgruber: Einleitung zu Akten der Reichskanzlei, Regierung Hitler 1933–1945, published by H.G. Hockerts, volume II: 1934/35, edited by F. Hartmansgruber, Munich 1999, p. LX, and Document 143, op cit, pp. 531–535, as well as the official grounds for the law set out by the Minister of Transport in: Die Strasse, issue 15, 1935, p. 565.

311 Law on Long Distance Goods Transport with Motorized Vehicles on 26 June 1935, in: RG, Teil I, 1935, pp. 788–793, as well as Ordinance for Implementation, in: RG, 1936, Teil I, p. 320–326. On the prior history of this law see note 5 of the editor Minuth on document No. 64, in: Akten der Reichskanzlei, Regierung Hitler 1933–1938, op cit, p. 320s. Until the enactment of this law the Brüning government's Overland Ordinance remained in force. For the further development of restrictions on private trucking companies up to 1945 see also Franz. W. Seidler: Fritz Todt, Bublies Verlag, Beltheim-Schnellbach 1986, pp.144–152.

have been satisfied with the weak position of the commercial road trucking industry, he continued the battle against trucks after his appointment as transport minister in 1937 and the re-incorporation of the *Reichsbahn* into his ministry as a state run company.[312] The number of trucks employed in private sector long distance trucking fell from its highest level of 13 400 in 1933 to 11 400 in 1937, thus falling even below the figure of 11 500 in 1932, notwithstanding the greater demand for transport services caused by rearmament.[313]

As Kopper pointed out, with regard to *autobahn* construction, transport policy was highly contradictory.[314] On the one hand, a network of 3000 km of high-speed roads was created in the years before 1939. On the other, truck transport by private trucking companies was drastically curtailed.[315] And also the strategies of the transport policy do not fit to the goal of re-armament and to the goal of a powerful railroad for transportation services in the war. Neither have there been incentives for the diffusion of trucks at private enterprises so the army could requisition trucks nor did the *autobahn* serve the deployment of troops. The following figure summarizes the contradictions of the NS transportation policy.

---

312 In a letter to Dorpmüller on 22 February 1938 Todt complained of the continued subsidization of cartage and the Reichsbahn's purchase of capacity on the market for commercial long distance transport, see BA, R5, volume 8125, sheet 84–89. Todt cites Order No K2.566 issued by the transport minister on 2 February 1938 on the emergency program to remedy the desperate situation in the long distance transport industry, see Die Strasse, issue 5, 1938, p. 161. On obstacles to trucking companies see also Schütz and Gruber, op cit, p. 139s.

313 W. Scholz: Die Entwicklung des Kraftfahrzeugbestandes im gewerblichen Güterfernverkehr, in: Die Strasse, issue 4, 1938, pp. 106–108, here p. 106.

314 Kopper 1995.

315 See also Oswald Lutz: 3000 Kilometer Reichsautobahn für den gewerblichen Kraftverkehr, in: Die Strasse, 1938, issue 24, p. 772.

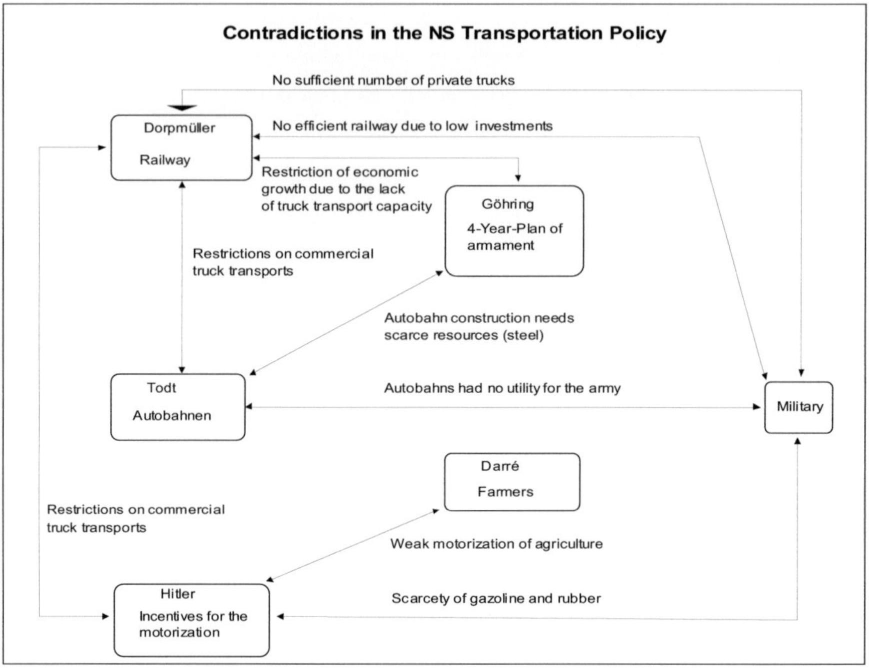

Image 24: Contradictions in the NS Transportation Policy

11.7 The Visions of Future use of the Autobahn

One factor, which decisively influenced the autobahn project, but which evades a formal discussion of goals, is Hitler's megalomania. Hitler predicted the shift in traffic away from the railway to *autobahn* and air transport networks with astounding precision. In an analogy to Henry Ford's T-model, some 15 million of which were produced between 1908 and 1927, he also recognized that German industry was not meeting the demand for a standardized entry-level model for mass motorization. However, an international comparison of German transport policy, raises the question of why *autobahns* were constructed in under-motorized Germany and not in Great Britain or the USA, which in per capita terms had a 15 times greater level of motorization than Germany. Evidently, there was initially no need for them even in the USA. Instead, in the 1920s multi-lane arterial roads led out of the large cities

Chicago, Detroit and New York.[316] Just like the construction projects planned by Speer at Hitler's order for the party congress stadium in Nuremberg and in the capital Berlin, the *autobahn* project also has a tendency to the gigantic.[317] While the architectural abnormality of the construction projects is immediately apparent, the gigantic nature of the *autobahn* is not obvious in the individual sections, but in the planned network of 7000 km and also in the bridge constructions.

11.8 The Autobahn and Local Traffic

The question of arterial roads leading from large cities is associated with the relationship between local and long distance traffic. So far, research on National Socialist *autobahn* construction has paid little attention to how it met the need for efficient local transport conditions for the trucking industry in spite of the fact that this had not initially been explicitly stated as a goal. The *autobahn* network was regarded purely as a long distance transport network. However, in a subsequent interpretation of its work in 1937, the Gezuvor referred to the goal of decentralization into the suburbs that was related to the *autobahn* project. *Autobahns* were supposed to "compensate for the urban planning predominance given to large railroad stations in the center of cities, the main railroad stations and main freight stations"[318] – a clear dig at the *Reichsbahn* that reveals the tensions between the two transport systems. However, this argument of decentralization into the suburbs is not very convincing, as here the need is less for *autobahns* that are a long way from the cities than the kind of multi-lane arterial roads out of towns that could already be found in the USA in the 1920s. The example of the Berlin *autobahn* circle shows that – apart from the western approach over the AVUS – the Berlin ring could only be reached from the city center over narrow country roads as no arterial roads were built.[319] Incidentally, this was to remain the case until 1995.

---

316 See Mark Foster: From Streetcar to Superhighways: American City Planners and Urban Transit 1900–1940, Philadelphia 1981. Neumann reports on multi lane arterial roads in San Francisco and of parkways in Chicago and Detroit, in: Hafarba-Mitteilungsblatt, issue 1, 1931, p.4. See also: Die neue Automobilhochstrasse zu Neuyork, in: ADAC-Motorwelt, 1930, issue 43/44, pp. 4–5 and Vilbig: Bericht über den VI. Internationalen Straßenkongreß in Washington, ADAC-Motorwelt, 1931, issue 1, p. 22.

317 Compare Joachim Fest: Speer – eine Biographie, Berlin 1999.

318 Gesellschaft zur Vorbereitung der Reichautobahnen (Ed.): Die Planungsarbeiten für die Reichsautobahnen – Zweieinhalb Jahre GEZUVOR, Berlin 1937, p. 128.

319 See the report, "Vom Potsdamer Platz zur Reichsautobahn Berlin–Stettin," in: Die Autobahn, 1936, p. 168s.

In fact, with the construction of the expressway, initially many local traffic connections were bolstered. The first construction measures carried out after 1933 were to link neighboring towns:[320] Frankfurt with Darmstadt, Mannheim with Heidelberg, Leipzig with Halle, Kassel with Göttingen, and Cologne with Düsseldorf.

Image 25: Autobahn Gateway to Mannheim

The Chiemsee *autobahn* was also built, making this recreational area more accessible from Munich. The following table summarized the openings of the first route segments from 1935 until 1936 in the Reich.[321] However, the support of local traffic was not a dominant principle of National Socialist *autobahn* construction, as will be shown later.

---

[320] This is shown by the map of the autobahn sections started from 1935 onward, see Die Strasse, issue 4, 1935, p. 104.

[321] Data from Statistisches Jahrbuch für das Deutsche Reich, 1936, p. 198.

Route	Lenght km	Date of Opening
Frankfurt a.M.–Darmstadt	22,0	19th of May 1935
München–Holzkirchen	25,0	29th of June 1935
Darmstadt–Mannheim (Heidelberg)	61,0	3th of October 1935
Holzkirchen–Weyarn	7,0	11th of January 1936
Leipzig–Halle	26,5	25th of April 1936
Köln–Düsseldorf	24,5	21st of May 1936
Weyarn–Rosenheim	33,0	24th of May 1936
Rosenheim–Siegsdorf	35,0	17th of August 1936

Table 6: Opening of the first route segments of the autobahn from 19th of May 1935 till 17th of August 1936

In this table it is interesting to note how swiftly the route Munich–Holzkirchen was completed, compared with the route Frankfurt–Darmstadt. The construction of the former route was begun just half a year later and was finished only a month after completion of the later. This underscores the great importance given to this route both by Hitler, who was personally involved in the route and visited its construction sites on various occasions, as well as by Todt.

12 Implementation and Legitimizing of the Autobahn Project

If one follows the decision-making process as documented in the Chancellery files, one has to conclude that Hitler was the decisive promoter of the *autobahn* project within the government apparatus. This fits to the intentional concept in the historical research on NS system.

On the other hand, according to the structural approach to the study of National Socialism[322] it can be assumed that he utilized the structures of the situation in 1933 and bundled them to form a single project:

- The need for job creation measures,
- The existence of the fully developed Hafraba plans for the section Frankfurt–Mannheim
- Players who were already attuned to one another from the attempts of the Brüning cabinet to implement the Hafraba plans.

In spite of this, Hitler had to overcome a great deal of resistance before he could commence *autobahn* construction with the founding of the *Gesellschaft Reichsautobahnen.* He then left implementation of the project to Todt. Due to the visionary character of the *autobahn* project, that contrasted sharply with the low level of motorization in Germany, there was resistance from a number of quarters and the lack of urgent transport or military goals for *autobahn* construction revealed a legitimacy deficit. Some simple comrades rejected *autobahn* construction on the grounds that it only served the motoring pleasure of the rich. The *Reichsbahn* opposed the *autobahn* project because it rejected anything that would promote competition from cars. Specialists from the ministries doubted the economic benefit and financing feasibility of the project. The controversy with the experts centered on whether funding would be better employed expanding the existing road network or constructing new *autobahns*. Hitler argued that the existing road system was designed for horse-drawn wagons and was therefore unsuitable both in terms of its substructure and road width for the burden posed by rapidly increasing modern motorized traffic, as demonstrated by the need for frequent repairs.

---

322 See Ian Kershaw: Der NS-Staat, Hamburg 1988.

Furthermore, allowing long distance traffic to pass through towns caused damage to buildings. Separating through-traffic from local traffic was vital for ensuring brisk automobile traffic on the roads.[323] Here he adopted those arguments for *autobahn* construction that had been repeatedly put forward since 1926 by the *autobahn* advocates belonging to the Hafraba association, published in the association's bulletin and also cited by other critics of the existing road system, for instance in Todt's memorandum of December 1932.[324]

Hitler was able to overcome the *Reichsbahn*'s opposition to the *autobahn* project when on 10 April 1933 he offered to let the head of the railroad, Dorpmüller, take over construction of the *autobahn*, thus involving the *Reichsbahn* in the *autobahn* project as a planner, financier and organizer.[325] By taking this approach, Hitler invalidated any objections that might have been raised by the Transport Ministry, which basically pursued railroad policy and thus reduced his cabinet's resistance to the *autobahn* project. At a meeting with Hitler on 1 June 1933, Dorpmüller agreed to assume responsibility for *autobahn* construction, requested the creation of the legal basis and offered 50 million RM in initial funding.[326] The legal basis was then created on 27 June 1933 with the law concerning the "*Errichtung eines Unternehmens Reichsautobahnen*," (Establishment of a Reichsautobahn company), §5 of which stipulated that the *Reichsbahn* would manage and represent the company. However with Todt, a road construction engineer, Hitler assigned a strong promoter of *autobahn* construction to the *Reichsbahn* as "General Inspector for German Roads". According to his biographer Seidler, he has to be considered a trucking lobbyist[327] and as such represented a counterweight to Dorpmüller. §8 and §11 of the law gave him the right to decide the routing of the *autobahns*.[328] After the successful initiation of *autobahn* construction, rights of control over the *Unternehmen Reichsautobahnen* were once more withdrawn from the *Reichsbahn* and transferred to Todt by a

---

323 K.-H. Minuth: Introduction to the files of the Reichskanzlei, Regierung Hitler 1933–1938, op. cit., p. XLIII as well as a discussion between the Chancellor and transport experts on the question of autobahn construction on 18 May 1933, op. cit., p. 463s., see also E. Schütz and E. Gruber, op. cit., pp. 19–22.

324 Straßenbau und Straßenverwaltung, BA, R 4601, volume 3001/A.

325 Hitler's discussion with the head of the railroad, Dorpmüller, on 10 April 1933, in: Akten der Reichskanzlei, Regierung Hitler 1933–1938, op. cit., p. 331. Parallel to the autobahn project, the Reichsbahn was able to realize new constructions and acquisitions worth 560 million RM as a work creation program of its own in the years 1933 and 1934.

326 Akten der Reichskanzlei, Regierung Hitler 1933–1938, op. cit., p. 543s. See also: A. Gottwaldt: Julius Dorpmüller, die Reichsbahn und die Autobahn, Berlin 1995, p. 41.

327 Seidler, p.146.

328 The press called him the dictator for road construction, see Völkischer Beobachter on 6 July 1933, p. 1.

decree issued on 23 January 1935. This was justified on the grounds that the *Reichsbahn* had not given *autobahn* construction the necessary priority nor proceeded with the necessary financial boldness.[329]

With the planning of the expressway, Todt abandoned the principles of railroad line-management; instead he linked the expressways to the concept of landscape aesthetics, that he had adapted from US parkways.[330] Todt's concept of an expressway shows a remarkable difference to railroad lines, which either run along river valleys or straightway through the landscape, cutting hills and crossing valleys over heaped up embankments. Because they are technical constructions which stand out in the landscape, travelers experience the train journey as a loss of landscape and estrangement from living nature, as Schivelbusch explains.[331] In contrast to the railroad layout, the course of the expressways was to be fitted into the landscape; neither steep embankments nor deep ditches were to divide the lanes from the surrounding landscape.[332] In addition, the car ride should be connected to an enjoyment of landscape.

By following a fundamentally different approach as railroad engineers have when designing traditional railroad lines, the takeover of railroad engineers for the planning offices of the *Reichautobahngesellschaft* confronted Todt with the problem of how to adapt the corporate culture of the *Reichsbahn* to the needs of the construction of the expressways. As a corrective measure, Todt assigned landscape lawyers to work with the main direction of constructions, who were to have an advisory function in fitting the expressway to the landscape.

---

329 Hartmannnsgruber, op. cit., p. LX, as well as the RG, Teil II, 1935, p. 37.

330 *Zeller 2002*, p. 163, see also R. Hoffmann, „Prachtstraßen (Noble Highways)“, in: Die Strasse, 1935, pp. 493–495, Wolfgang Singer, „Parkstraßen in den Vereinigten Staaten von Nordamerika“, in: Die Strasse, 1935, pp. 175–178, Bruno Wehner, „Die landschaftliche Ausgestaltung der nordamerikanischen Park- und Verkehrsstraßen“, in: Die Strasse, 1936, pp. 599–601. On the Merrit Parkway see *B. Radde*, The Merrit Parkway, Yale University Press 1996, also: Alexander Thewalt*:* Projekte der Autobahnfrühzeit im Ausland – Beispiele aus Italien und den USA, in: W. Wirth: Die Autobahn – Von der Idee zur Wirklichkeit, Köln 2005, pp. 63–76. Also: Bernd Kreuzer*:* Tempo 130, op. cit., pp. 13–17.

331 Wolfgang Schivelbusch*:* Die Eisenbahnnreise, Frankfurt 2000, p. 27.

332 In his letter to Seifert about the route Munich–Holzkirchen from the 23 November 1933 Todt expresses the aim that the expressway should not seem an "artificial lane (railroad embankment)", but rather the driver should "sense a closeness to nature in these magnificent woods", ADM, NL 133/056.

To conceal the lack of specific transport policy objectives, the *autobahn* project was not only portrayed as a measure to create employment and promote motorized transport. Instead, propagandists shifted the discussion to a whole series of other aspects:

- The project would create a giant work for the future of transport.
- It would build a monument to National Socialism that would stand for centuries.
- "Adolf Hitler's roads" would be built.
- Principles of landscape aesthetics would be realized. Autobahns would be adapted to the landscape. At the same time, the landscape was to be cared for.[333]
- A pseudo-ideology was adopted in order to reconcile technology and nature.
- The aspect of recreation was combined with the autobahn. The peculiar idea of "a*utowandern,*" (road touring) was introduced, that was supposed to combine enjoyment of the landscape with driving through the landscape on the autobahn.[334]

The particles of ideology listed here were spread among the population in countless inaugural speeches, films, exhibitions, publications and, above all, in newspaper reports in the *gleich-geschaltete* press.[335]

The bridges of the *autobahn* project were a special aspect upon which the Nazi party wished to express its power and its plans for a long-lasting empire. Monumental a*utobahn* bridges[336] were embedded in the concept of landscape aesthetics[337] and were supposed to stand out as

---

333 Shand 1984, op. cit., interprets landscape aesthetics as a means of reconciling the modernity of the autobahn with the Nazi's völkisch ideology.

334 On "Autowandern" see Die Strasse, issues 3, 11, 14, 1936.

335 On the traveling exhibition "Die Strasse" see Die Autobahn, 1934, issue 8, and issue 10, p. 442 and issue 15, pp. 685–690. The traveling exhibition was shown in Munich, Berlin, Essen and Braunschweig. At the opening of one section near Dresden on 26 June 1937 Hitler emphasised the future of transport: "In a few decades there will be massive traffic on these roads, in which the entire population will participate. Millions of our Volkswagens....will roll over these roads." See Kasseler Post, 27 June 1937, Stadtarchiv Kassel.

336 See Karl Schaechterle: Die Gestaltung der Wegüberführungen über die Reichsautobahn, in: Die Strasse, issue 8, 1934, p. 239s. Further: numbers 8 and 23 of the journal "Die Strasse," published by Todt, volume 1935, that focused on Autobahn bridges, as well as numbers from later years A study of the bridges was made by R. Stommer: Triumph der Technik – Autobahnbrücken zwischen Ingenieuraufgabe und Kulturdenkmal, in: R. Stommer (Ed.): Reichsautobahnen – Pyramiden des Dritten Reiches, Marburg 1982. See also the study on bridges in E. Schütz and E. Gruber, Mythos Reichsautobahn, Berlin 1996, 2nd edition 2000.

337 Also in the USA bridges played a role in the parkways. See the ornaments of Jugendstil at the bridges of the Merrit Parkway, in: Bruce RADDE: The Merrit Parkway, Yale University Press 1996

symbols of eternity, thus increasing the prestige of the Third Reich.[338] Image 26 depicts a bridge crossing the Berlin *autobahn* circle.[339]

Image 26: Bridge crossing the *Autobahn* belt around Berlin

In this regard, it is hard to avoid the impression that monumentality was imposed with no regard for construction costs. Landscape aesthetics and monumentality were prime goals that took priority over economic efficiency and safe driving. But ironically, the monumental aspect of the bridges hid some severe technical defects in the construction of the bridges, which have been discovered by researchers in later times.[340] The Nazis had speeded up the construction of the bridges at the expense of their stability and of road safety to cope with the goal set up by their propaganda to complete 1000 km *autobahn* every year from 1936 on. And the monumentality was short-lived, because in 1945, in the last days of war, it was ironically the German troops themselves that, in the territory of the later Federal Republic alone, blew up 517 expressway bridges[341] to impede the advance of the allied troops, thus massively disrupting the expressway network. This, not least, led to a severe crisis of the German

338 The monument aspect was emphasized several times. In 1934 the GEZUVOR (society to prepare the autobahn) section head Becker provided some unintentional comedy when he compared the construction of the autobahns with that of the pyramids in Egypt, see Becker: "Von der HAFRABA zur Reichsautobahn“ in: Kasseler Post on 18 March 1934, p. 3, UBK. At the inaugural ceremony of a section near Dresden on 26 June 1937, Hitler said that the autobahn would still be standing in a thousand years, see Kasseler Post, 27 June 1937, Stadtarchiv Kassel.

339 Usinger, Carl: Der Berliner Ring der Reichsautobahn, in: Die Strasse, issue 6, 1937, after p. 148

340 The bridge Wommen in the section Bad Hersfeld–Erfurt had sunk into the ground by 10 cm and had to be reinforced with concrete after the reunification of Germany in the 1990s. I owe this hint to my colleague Wolfgang Jäger, Münster. The bridge over the river Werra at Hannover'sch Münden in the section Göttingen–Kassel did not properly smooth the steep ascent to the hill at the end of the bridge, so that buses scraped street with their bodies.

341 Bundesminister für Verkehr (ed.): Die Verkehrspolitik in der Bundesrepublik Deutschland 1949–1953, Dortmund 1953, p. 87.

transport economy from 1945 to 1948, and drastically restricted supply to industry and population.

However, *autobahn* bridges had to be different from bridges used in the railway network to underline the modernity of the concept of the *autobahn* and the deep rivalry between *autobahn* and railway. In the first period of railway building in the 19th century, bridges were constructed out of stones and concrete in the form of the stone arch viaducts. But after 1880, the time where steel became a relatively cheap material, the railway bridges were mostly constructed out of rods of steel enabling a greater span (trussed girder bridges and arches through bridges). And passing the rods, the passenger in a railway wagon can realize that he is crossing a bridge, cf. Image 27.

Image 27: The Railway Bridge crossing the River Rhine in Cologne in 1910[342]

The modern concept of *autobahn* bridges that the Nazis had implemented is a construction with steel girders crossing a valley.[343] The driver does not realize that he crosses a bridge and he has the impression of flying over a valley, cf. Image 28.

---

342 Archive of Deutsches Museum Munich, image no. 10283.

343 Schaechterle: Reichsautobahnen und Stahlbrückenbau, in: Die Autobahn, issue 19, 1934, pp.909–911.

Image 28: Bridge Crossing Valley Aula close to Bad Hersfeld[344]

This feeling of flight was expressed by the bridge architect Paul Bonatz,[345] who wrote seven articles for Todt's journal. But the use of steel girder bridges was stopped at the beginning of the armament period, 1936–1939, to save the scarce steel and bridge designers returned to the form of the stone arch viaducts.[346] However, to construct monumental bridges is not a uniform principle, even in sections which are directed at the principle of landscaping. In the section Frankfurt a.M.–Kassel, one sees many bridges constructed in a functional manner (cf. Image 41).

Because the goal set by propaganda was to complete 1000 kilometers of expressway annually from 1936 on, the project stood under great time pressure. From 1936 to 1938, the targets were achieved on time and the completion dates were greatly celebrated. However, from mid 1938 fortification works on the western border constrained the building of the expressway, since construction capacity had to be removed. The construction target of 3000 km in 1938 was only achieved in time through provisionally laid lanes – partly single-lanes – on the last

---

344 Helmut Woite: Brücken der Reichsautobahn Göttingen–Kassel–Bad Hersfeld, in: Die Strasse, issue 15, 1939, p. 483.

345 Dr. Todt und seine Reichsautobahnen, in: Die Baukunst, March 1942, p.51, cited according to Franz Seidler: Fritz Todt, Bublies Verlag, Beltheim-Schnellbach 2000, p. 112.

346 Schaper, Gottwalt: Stahlsparende Baustoffe und Bauweisen im Sinne des Vierjahresplans, in: Die Strasse, p. 465–568, issue 16, 1937.

kilometers. Of the 3000 kilometers, 9.8% were single-lanes. In 1939 construction performance sank to 255 km, due to war preparations, but bounced back to 436 km in 1940, and the important gap in the route Hannover–Dortmund was closed, so that the route Berlin–Cologne could be driven non-stop. Until summer 1943 the network had grown to a total of 3858 km, of which 14.3% were single-lanes.[347] A map of the network, as it has been established till 1939, shows the following image.[348]

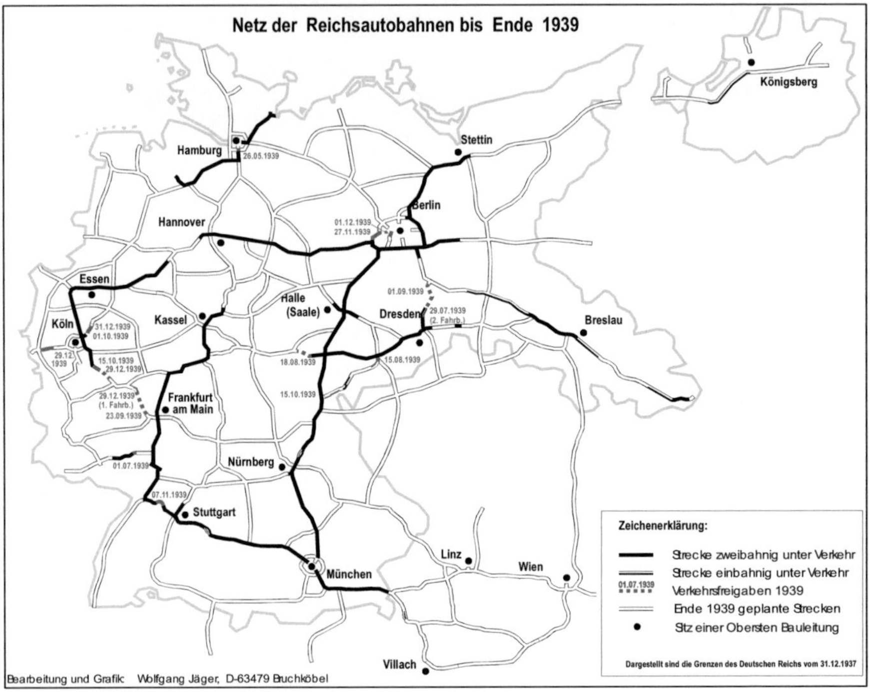

Image 29: Network of the Autobahn up to 1939

347 Jäger and Schwell, Verkehrsfreigaben von Autobahnen in den Jahren 1933 bis 1945, in: Strasse und Autobahn, 2000, issue 8., p. 488.

348 Ibidem.

13 The Lack of Planning Guidelines for GEZUVOR

On 18 August 1933, the Hafraba was dismantled and transformed into the *Gezuvor – Gesellschaft zur Vorbereitung der Reichsautobahnen e.V.* (GEZUVOR – Society for Preparation of the Reich Expressways).[349] Most former board members were taken over in the new society. The résumés and portrait photographs of the new chairmen were published in issue 9 of *Die Autobahn* of 1933.[350] The society was relocated from Frankfurt to Berlin and divided into eleven regional sections with merely regional responsibility. The section Hannover–Kassel was headed by the *Landesoberbaurat* Becker (State Senior Building Officer). This regionalization of the Gezuvor caused Todt's power to grow. He was given the exclusive authority over the entire network. In his lecture in Frankfurt before high-ranking state representatives on the occasion of the transformation of the Hafraba into the Gezuvor, Todt warned the organization to confine themselves to subsidiary issues, but to refrain from central planning.[351]

As with the Hafraba, the Gezuvor was not part of a branch of the transport administration and its formal character was that of a private association. The work of Gezuvor was financed by the municipalities and chambers of industry and commerce as institutional members of the Gezuvor. The number of members grew from 300 in 1933 to 3000 in 1934, as many towns became member (forced by the Prussian President Göring) along with companies from the road construction industry.[352]

Although Willy Hof – and not Fritz Todt – was the formal head of the Gezuvor, Todt could entrust them with the task of planning various routes. To carry out the planning, eleven regional sections that were created were run with a strict hand by Todt. Some of the sections

---

349 Die Autobahn, issue 9, 1933, pp. 7–9.

350 Direktor Willy Hof, Artur Zierau (senior executive officer of Baden's road and water construction authority), Niemeyer (head of the municipal planning and building control office in Frankfurt) and Nienhaus (mayor of Heidelberg) were members of the NS-party at the time of August 1933, in the place cited.

351 Städtisches Anzeigenblatt Frankfurt a.M. on 26.8.1933, p. 378f, ISG.

352 The member's list as on 28 October 1933 in: BA volume R4601, volume 1263, sheet 86–90. For the new members see Die Autobahn, Vol. 1934, passim.

handed in their reports at 14-day intervals.[353] In 1933 the Gezuvor comprised over 300 experts who, showing tremendous commitment, worked late into the night drawing up the plans. What makes the Gezuvor's activities so remarkable is that most of these experts worked in a voluntary capacity. Todt's dictatorial powers allowed him to exploit the Gezuvor with cold calculation. Within a short period, and at no cost, he obtained plans for what he called the basic network. This road system was just under 7000 km long and was documented in 387 volumes with sections of around 30 km.[354] None of these volumes can be found in the archives of the Federal Republic of Germany today.

By the end of 1933, the Gezuvor had submitted a total of five reports about preliminary project work to the Chancellor through Todt. In 1934 a further 20 reports were submitted.[355] The fact that the Gezuvor addressed its reports to the Chancellor shows that Hitler had followed planning progress. The Gezuvor fell into a planning euphoria and wanted to expand the basic network from 7000 km to 11 000 km, prompting Todt to admonish that expansions to the basic network should be postponed and work concentrated on urgently needed sections instead.[356] Interestingly, the Gezuvor initially planned without uniform guidelines leading Schönleben, head of Todt' department, to make an anxious enquiry on 12' October 1933 asking the Gezuvor which planning guidelines it used.[357] The "Guidelines for Autobahn Planning"[358] were only laid down in a decree for the Gezuvor and the *Reichsautobahn* Company by Todt on 12 December 1933.

The Rhine-Main Gezuvor section Rhine-Main had, in short reports, presented in detail the economic and transport policy relevance of five subsections of the expressway network.[359]

---

353 BA volume R4601, volume 1263.

354 Gesellschaft zur Vorbereitung der Reichsautobahnen (publisher): op. cit., p. 133. In the introduction, Minister Kerrl cynically referred to the work carried out by the GEZUVOR free of charge as "the members' willingness to make sacrifices." See also the minutes of the last meeting of the GEZUVOR on 19 December 1935, op. cit., pp.134–137.

355 BA volume R4602, volume 600 and volume 1.

356 Todt to the GEZUVOR on 4 June 1934, BA volume R4601, volume 1263, sheet 19, 51, 52. At a ceremony marking the presentation of 2000km of autobahn on 17 December 1937 with 2000 construction workers, Hitler talked about an expansion to 12 000km, see Die Strasse, issue, 1, 1938, p. 2.

357 in: BA R 4601, volume 1263, sheet 105.

358 in: BA R 4601, volume 1263, Sheet 65–70.

359 Magistratsakten 6289, ISG.

These surveys were made for the sections Frankfurt a. M.–Gießen–Bad Hersfeld (25 April 1934), Frankfurt a. M.–Aschaffenburg (25 April 1934), Frankfurt a. M.–Trier (15 December 1934), Asbach–Gießen (15 December 1934), and Frankfurt a. M.–Köln (2 December 1933).

Here, by way of example, we will expound upon the explanatory report on the section Frankfurt a. M.–Gießen–Hersfeld. Following the geotechnical approach of Fritz Stück, an urban planner from Kassel (cf. section 20),[360] the report arrays the section Frankfurt a. M.–Hersfeld both in a north-south axis (North Sea–Mediterranean), and a west-east axis (Paris–Frankfurt a. M.–Berlin). In a further step the substantiation of the connection of the central German industrial area with the industrial area of Rhine–Main–Neckar–Saarland, is worked out in detail. The importance of the route for tourism is emphasized, as it connected the well-known spas Bad Soden, Bad Homburg and Bad Nauheim in the Taunus region, as well as numerous other health resorts. To increase the appeal of the world-famous spas of Taunus, a 10 km long racetrack, between kilometer 23 and kilometer 33, was proposed. The preliminary work for this racetrack was already begun, in accord with the competent authority, the president of the *Deutschen Automobilclub* in Berlin.

The economic importance of the route was highlighted by explaining that it would bring economic restructuring to regions which until then had been very much neglected, the Schwalm for example, and it could also provide a connection to the Fulda Valley. The upper Hessian districts boast the following industries that process natural resources or manufacture semi-finished or finished products:

The marble and slate quarries of Lahn, the basalt and cobblestone industries in Vogelsberg and in the district of Lahn; the furniture and wood-processing industries in Taunus and Vogelsberg; the papermaking industry in Lahn and Fulda; the textile and canvas factories in Fulda and the Schwalm area; the luxury-food and cigar industries in Fulda and the Lahn area, and the soap, candle, wax and perfume industries in the areas of Künzig, Fulda and Lahn. The expressway was ascribed an important role in the boosting of iron and ore mining in Oberhessen. Besides smelting and foundry plants, the machine, oven and stove manufacturers

[360] See section 20.

– among them the „Buderus'schen Eisenwerke Wetzlar" with their world-renowned products – must also be mentioned. The transport needs, thus, are quite general, but lack empirically verifiable data.

According to the Gezuvor files that have been found at the Federal Archive[361], a large number of different routes were planned, especially in central Germany, that were handled by the Hanover–Kassel section. The Hanover–Kassel section was run by *Landesbaurat* Becker, who had previously been a member of the Hafraba board, and who was known as the author of numerous articles about the planting of vegetation on roadsides. Responsibility for *autobahn* planning was withdrawn from the Gezuvor at the end of 1935 after the planning of the basic network had been completed,[362] and Becker ceased to play a role in *autobahn* development. Todt's many requests for routes threatened to overwhelm the sections with work. According to the Hanover–Kassel section's report on 14 April 1934, the following routes were planned:[363]

- Göttingen–Kassel–Fulda–Würzburg
- Eisenach–Bad Hersfeld–Gießen
- Kassel–Eisenach–Bamberg
- Emden–Münster–Kassel

According to the Hanover–Kassel section's report on 30 May 1934, the following additional routes were also taken on.[364]

- Kassel–Gotha
- Kassel–Cologne

The fifth report on Gezuvor's preliminary project work was dated 20 December 1933 and also mentions the Eisenach–Fulda–Frankfurt route.[365]

---

361 BA volume R4602, volume 600 and volume 1, and R4601, volume 1476 and volume 1263. In the archives of the German states there may exist additional files on the GEZUVOR, e.g. Thüringisches Hauptstaatsarchiv Weimar, C233, C234, C235.

362 This competence was withdrawn at an extraordinary meeting of members on 27 July 1935, see Die Strasse, issue 15, 1935, p. 558s.

363 BA R4602, volume 600.

364 Op. cit.

365 Op. cit.

The interesting thing about these plans is that in the area Hannover–Kassel–Frankfurt, they diverged from the original Hafraba-route Göttingen–Kassel–Marburg–Gießen–Frankfurt (cf. Image 43).[366] After the first design – the Hafraba-line – which was completed in 1926/1927, the route was again optimized in the years 1930/1931 in consultation with the cities Göttingen, Hannoversch Münden and Kassel. Thus the route Göttingen–Kassel originally passed through the western suburbs of Kassel, intersecting Wilhelmshöhe and crossing the Weser just north of Veckerhagen, and then, north of Göttingen, pivoting back to the present day route to Hanover. In agreement with the three cities, a routing was chosen that came very close to the one realized in the years 1935 to 1937, fixing a maximum gradient of 5%.[367]

Instead of the Hafraba-route, Todt determined by letter on 18 September 1933, that the north-south-route Göttingen–Kassel–Fulda be planned.[368] The term "Hafraba-route" is no longer mentioned in the Gezuvor files at all.[369] In a retrospective interpretation of decision-making procedures that led to route-planning changes, in 1962, former *Oberregierungsbaurat* (state construction engineer) Karl Denners suggested these occurred because Todt gave priority to the route Hamburg–Munich over Hamburg–Basel.[370] At first glance, this seems plausible because it would have meant the planning of a north-south route crossing the entire Reich and connecting two urban agglomerations. The inclusion of Munich in the north-south axis is also logical if one considers the importance of this city for the National Socialist Movement and also the many years that Todt worked there. However, in the files of the Gezuvor and the *Oberste Bauleitung* (main construction directorate) in Kassel, only references to the

---

366 R. Otzen: Die Autostrasse Hansestädte–Frankfurt–Basel. Erläuterungsbericht zum Entwurf der Autostrasse, Hannover 1927, Schrift Nr. 4, p. 10s.

367 Landesoberbaurat Kurt Becker: Göttingen–Hann. Münden–Kassel, in: Mitteilungsblatt Nr. 2, 1931, pp. 1–4 as well as Folckert Lüken-Isberner: Fritz Stück and the urban planning discussion in Kassel during the Weimar Republic, in: Hessisches Jahrbuch für Landesgeschichte, vol. 40, 1990, pp. 219–240, here 237–239. On the unhappy role Stück played as the landscape lawyer of the construction direction in Kassel, Zeller, 2002, p 100s.

368 Activity report of the section Hannover–Kassel on 30 December 1933, BA 4602/600.

369 In the press of Kassel, the Hafraba term was still used, see the lecture of the state senior building officer Dr. Kurt Becker "Von der Hafraba zur Autobahn" ("from Hafraba to the expressway") in: Kasseler Post on 18.3.1934, UBK and „Kurhessen im deutschen Raum – Kassel im Schnittpunkt der Reichsautobahnen", in: Kurhessische Landeszeitung on 2 May1936, UBK. Todt's magazine "Die Strasse" refers only once to the Hafraba, when reporting on the opening of the route Frankfurt–Darmstadt in May 1935, see K.Kaftan: „Hafraba" – Ein Rückblick, in: Die Strasse, issue 9, 1. Maiheft, 1935, p. 321s.

370 Karl Denner: Planung und Entwurf der Strecke Kassel–Frankfurt–Mannheim, pp. 97–100, here p. 99, in: Bundesminister für Verkehr (ed.): Hafraba – Bundesautobahnen Hansestädte–Frankfurt–Basel, Bau Verlag, Wiesbaden, 1962.

Göttingen–Kassel–Fulda route or the Göttingen–Kassel–Fulda–Würzburg route can be found. According to the Gezuvor plans, the route would have been continued from Würzburg over Augsburg to Munich.[371] This route was called "the cities route". In a published presentation about expressways completed by 15 December 1938[372] neither a route Würzburg–Munich nor a route Hamburg–Hannover–Göttingen is discernible and it appears that the north-south-route did not have a high priority.[373] In the completed network the central position of Berlin in the grid is immediately visible. Routes were built from Berlin to the following cities:

- Stettin,
- Magdeburg,
- Leipzig,
- Dresden,
- Breslau and
- Frankfurt an der Oder.

Although a catalogue of planning criteria was presented by the former Gezuvor planners in 1937,[374] this must be viewed as a subsequent rationalization because early Gezuvor planning did not follow any rational transport policy criteria for their route planning.[375] The lack of planning criteria greatly challenged Becker, of the Hanover–Kassel section, who was required by Todt to include a chapter on the economic impact of the route in his report. Becker, however, had no expertise in this area.[376]

Even today, we do not know which criteria and priorities formed the basis of Todt's decision-making when, at last, he gave order for the construction of specific sections from among the vast array of planning options. What emerges from articles in the journal "*Die Strasse*" is that while individual activities such as the construction of a junction for the Hamm–Kassel route[377]

---

371 Gesellschaft zur Vorbereitung der Reichautobahnen (ed.), in the place cited, p. 122.

372 Reichsautobahnkarte, in: Die Strasse, issue 24, after p. 765. Schwell and Jäger 2000.

373 In 1938 Todt assigned a preliminary project for the route Würzburg–Augsburg to be executed, see Die Strasse, issue 11, 1938, p. 362, and specified details for the route Hamburg–Kassel, that between Soltau and Gandersheim was to be laid in two parallel strips, see Die Strasse, issue 14, 1937, p. 419.

374 Gesellschaft zur Vorbereitung der Reichautobahnen (publisher), op. cit., p.127s.

375 Lärmer, op. cit., p. 94.

376 Progress report of the Hannover Kassel Section from 13 November 1933, op. cit..

377 See short reports, in: Die Strasse, issue 24, 1937, p.744.

or preliminary work for the Cologne–Kassel route[378] were started, the public debate that was conducted in the *gleichgeschaltete* press did nothing to clarify the overall concept. Todt regarded the Gezuvor's plans as mere preliminary work that he could change at will.[379] Indeed, the lack of a public debate on the selection of certain routes over others, or their cost-benefit relation which might have served as a democratic counterpart to the dynamics of the nation-wide mega project, left Todt with dictatorial powers. He could make irrational decisions or serve special interests unchecked.

The NS-expressway project, by taking over the Hafraba plans, also took over its visionary and utopian character, which in view of the low car density in Germany was particularly obvious and grotesquely heightened. This was linked to the lack of rational planning criteria of transport and economic policies, on the basis of which the STUFA had traced the network in 1926:[380] It had identified economic areas which produced agricultural products, natural resources or industrial goods, and linked these, as source and target areas, with a road network. The starting point of their planning was hence the expected traffic of goods and people. This planning rationality was however completely abandoned during the NS-era. This policy stands in stark contrast to the construction of the US expressway network. As Seely's analysis has demonstrated, the American expressway network was constructed stepwise, according to traffic needs, and followed strict cost-benefits calculations. He emphasizes that American engineers, because of low traffic density on country roads, saw no need for four-lane expressways except in a few regions.[381]

I would now like to discuss three planning criteria which could have been used and their relation to the layout of the autobahn network. It was not explicitly stated that **traffic density** in the different cities should be a criteria for the planning of routes. With one motor vehicle (including motorcycles) per 24 inhabitants, Munich had the highest vehicle density in

---

378 Op. cit.

379 F. Todt: Die Planung des Reichsautobahnnetzes, in: Die Strasse, issue 7, 1937, p. 176.

380 Rappaport: Das Netz der Kraftwagenstrassen Deutschlands, in: Verkehrstechnik, issue 18, 1926, pp. 285–288.

381 Bruce E. Seely: Building the American Highway System – Engineers as Policy Makers, Philadelphia 1978, Bruce E. Seely: Der amerikanische Blick auf die deutschen Autobahnen. Deutsche und amerikanische Autobahnbauer 1930–1965, In: Werkstatt Geschichte, issue 21, Hamburg 1998, pp. 13–15.

Germany (compared to Berlin with 38 inhabitants).[382] But in fact, Munich's plan for the *autobahn* to Chiemsee were primarily based on the expected Sunday day-trip traffic and Hitler personally laid down the details of the route. According to Kaftan, too many visitors from Munich were placing an excessive burden on the destination Starnberger See, and alternatives were sought in consultation with the construction industry (see section 21).[383] In 1937, traffic counts carried out on the autobahn to Chiemssee clearly showed the peaks caused by excursion traffic on Sundays.[384] A second criterion on which autobahn routing could have been based was the **intensity of freight traffic.** According to statistics from the GEDEKRA *Reichsverband der Kraftverkehre*, the strongest incidence of truck traffic in July 1933 was from Berlin to Cologne.[385]

Another criterion for autobahn routing could have been the **accessibility** to and from Berlin. However, if one looks at the connection Frankfurt a.M.–Berlin, it becomes apparent that planning and construction in Hesse deviated from this. The Frankfurt–Fulda–Eisenach stretch mentioned in the Gezuvor's fifth report about the preliminary work dated 20 December 1933, is the most direct link from Frankfurt a.M. to Berlin. However, this route was abandoned in subsequent Gezuvor plans and replaced by the less direct Frankfurt–Gießen–Bad Hersfeld–Eisenach route that still exists today.

Apart from easing local traffic in regions with a noticeably high traffic volume in the years before 1933 (though not mentioned in the files as a planning objective), the routing of the *autobahn* network conforms to no recognisable transport criteria. All that can be detected are formal aspects such as a layout for three lines to run from east to west and two lines from

---

382 Statistisches Jahrbuch für das deutsche Reich, 1933, p. 159.

383 Kaftan, op. cit., p. 153, and Anton Woerner: Die Reichautobahn München–Salzburg, in: Die Autobahn, issue 9, 1933, pp. 10–12.

384 For the data see Forschungsgesellschaft für das Strassenwesen (publisher): Strassenbau-Jahrbuch 1937–1938, Berlin, Verlag Volk und Reich, 1938, pp. 29–42, here p. 32s.

385 Konrad Mellerovicz: Autobahnen und Kraftverkehrswirtschaft, in: Die Autobahn, issue 9, 1934, p. 370.

north to south.[386] Only later, when the autobahn system had already achieved a significant dimension, were systematic traffic counts carried out.[387]

[386] E. Schönleben: Das Netz der Reichsautobahnen, in: Die Strasse, issue 19, 1936, p. 614 describes some 7000 km of the basic network.

[387] See: Der Kraftverkehr auf Reichsautobahnen, Reichs- und Landstrassen im Dritten Reich, published by the Generalinspekteur für das deutsche Strassenwesen, Berlin 1939.

14 The Start of the NS-*Autobahn* Project 1933–1934 in Frankfurt a. M.

In the spring and summer of 1933, decisions and planning steps were hastened, leading to the start of construction work. At the general meeting of the Hafraba on 23 May 1933, the managing director, Willy Hof, reported on his talks with Hitler and the latter's intention to extend the Hafraba-concept into a network to span the entire nation.[388] Kaftan's memoir suggests that Hafraba's expressway designs from Golder and Otzen were simply adopted by Hitler.[389] However he doesn't mention that Hafraba's board vehemently opposed the extension of the north-south route into a network.

At a general meeting in Frankfurt called on 23 May 1933, the Hafraba was "brought to toe the party line". The meeting was chaired by the new NS-Mayor Dr. Friedrich Krebs, who had ousted Dr. Ludwig Landmann from office and taken over in the magistrate meeting on 13 March 1933.[390] The members who had fallen into political disfavor did not attend. It was euphemistically imparted that their attendance was prevented by their "exercising the duties of their current function".[391] The local assemblage of *Kampfbund der deutschen Architekten und Ingenieure* (League of Struggle of German Architects and Engineers) from the wider Frankfurt area – which was headed by Fritz Todt – asked Krebs, in a letter on 13 June 1933, to subject the Hafraba plans to a thorough examination before making them the basis of the regime's employment program because the Hafraba "had merely served the economic interests

---

388 Protokoll der Mitgliederversammlung, Die Autobahn, issue 6, 1933, p. 7; Repgen, Konrad (Ed.) (1988): Akten der Reichskanzlei, Regierung Hitler 1933–1938, part I, 1933/34, volume 1, Boppard am Rhein, p. 305, as well as the report of the managing director of Hafraba e.V., Hof, about his meeting with the Reich Chancellor about the planning and financing of the expressway, 6 April 1933, op. cit., p. 308. The editor of the files of the Reich chancery, Minuth, has drawn attention to the fact that no files can be found about Hof's meeting with Hitler on the 23 March1933, as cited by Kaftan, and which Hof describes in his report as "first reception and talk", p. 306.

389 Kaftan, op. cit., pp. 165–168.

390 Städtisches Anzeigenblatt der Stadt Frankfurt a. M., Nr. 12, 1933, p. 148. See also Bettina Tüffers: Der Braune Magistrat, Personalstruktur und Machtverhältnisse in der Frankfurter Stadtregierung 1933–1945, Frankfurt 2004 (Studien zur Frankfurter Geschichte Bd. 54).

391 Protocol of a membership meeting, Die Autobahn, issue 6, 1933, p. 7.

of capitalist and Jews". Not wanting to endanger the rapid start of construction, the magistrate of Frankfurt rejected this appeal.[392]

On 14 June 1933 Todt authorized, at a short notice, the start of constructions for the route section Frankfurt a. M.–Darmstadt – the very first of the entire NS-expressway project.[393] By July 1933, Hafraba's chairman Adolf Zierau could declare triumphantly in the *Badischen Beobachter* that the priority Hafraba had given the planning of the route Frankfurt–Mannheim had been acknowledged by the *Reichsbahn* as well as by Todt.[394] This stance changed, however, when Todt claimed Hitler's authorship for the design of the expressway and threatened Kaftan by way of letter to comply with this policy.[395]

The route Frankfurt–Darmstadt was designed as a grand avenue with a magnificent width of 23 meters. Two lanes of 7.50 meters in each direction, a five meter-wide middle strip and shoulders of 1,50 meters width added to this imposing number. Todt clearly exceeded the 20.50 meters planned by Hafraba. It is remarkable that there are few publications or other sources that address the remarkable width of the newly built expressway. Whereas the expressway network and the layout of the route were extensively discussed in the press and professional publications, there are no specific references to the width at the beginning of the planning. Even Todt's edict „*Richtlinien für die Entwurfsbearbeitung von Autobahnen*" (Guidelines to Draft Processing of Expressways) from 12 December 1933 fails to broach the subject; instead it features radii for curves and gradients.[396] The width is first recorded in circular number seventeen of the Gezuvor' Berlin headquarters where it is defined as standard.[397]

In contemporary publications, and also in literature, it was pointed out again and again that the circumstance that the Frankfurt–Darmstadt route followed Hafraba plans allowed it to be

392 ISG, Magistratsakten MA6289. The mayor of Frankfurt, Krebs, was not identical to the chairman of Hafraba.

393 Völkischer Beobachter on 25 June 1933, p. 2, edition A., ISG.

394 Badischer Beobachter on 19 July 1933, Stadtarchiv Karlsruhe, 1/H-1596.

395 Kaftan in Brücke und Strasse, 1952, issue 4, p. 90. Todt wanted his letter to Kaftan to be understood as a warning, as Kaftan wrote.

396 BA 4601/5001.

397 Thüringisches Hauptstaatsarchiv Weimar, C234.

realized so swiftly. This statement will not be questioned here, instead it will be pointed out that there has been – to this day – no evidence from the archives to substantiate this assertion. The four supplemental volumes to the Hafraba application from August 1932 to the *Reichsfinanzministerium* (Reich Finance Ministry), in which the planned route Frankfurt–Darmstadt is recorded in detail, are neither to be found in the *Bundesarchiv* (Federal Archives) nor in the *Hessischen Hauptstaatarchiv* (Hessian Main State Archive). It is also unknown if the route Frankfurt–Darmstadt, as planned and engineered by Todt, adhered exactly to the routing of the Hafraba plans or if there were divergencies.

At the start of the surveying work, according to press reports, the plans worked out by the Hafraba for this route could largely be adopted and implemented. The Hafraba had already recorded groundwater conditions for the route and assessed drainage and sewage, as the press conceded frankly.[398] Compared to the Hafraba's original plan (see Image 43), the route was laid further east – in the area of Mörfelden. According to a press release by Dorpmüller, the route was to become "a paradise for car drivers".[399]

It is remarkable that the visionary and utopian character – in view of low car-density in Germany – of the Hafraba-expressway plans was also transferred to – and grotesquely heightened – in, the NS-expressway project. This stands in marked contrast to the development of the American highway network which was extended according to growing demand and followed strict cost-benefit calculations.[400] The express highway system built in the USA followed the German expressway example. It was not constructed until the 1950s and, according to Rose's appraisal, came as a reaction to heavy interurban traffic – decades too late.[401] In the 1950s in Europe, this idealistic path – the preceding of transport needs with an expressway network – was again seen in Holland where it also required political justification.[402]

---

398 Badische Presse on 3 July 1933, ibidem, as well as Frankfurter Nachrichten on 30 June 1933, ibidem.

399 Frankfurter Nachrichten on 30 June 1933, ISG, Badische Presse on 30 July 1933, Stadtarchiv Karlsruhe, 1/H-1596.

400 Seely: Der amerikanische Blick auf die deutschen Autobahnen, op. cit.

401 Mark M. Rose: Interstate – Express Highway Politics 1939–1989, Tennessee UP, second rev. edition, 1990, p. 92.

402 Mom, op. cit.

A great spectacle attended the official beginning of the construction work on 23 September 1933. Unemployed men from the labor offices gathered for a ceremonial act at the stock exchange in Frankfurt where the Gauleiter Jakob Sprenger handed out simple shovels.[403] The troop of unemployed marched like soldiers with shouldered shovels, headed by Todt and Gauleiter Sprenger, straight through the city to Schwanheim on the southern banks of the Main, cf. Image 30.

Image 30: March of workers through Frankfurt a. M. across the Scharnhorststraße (today Baseler Straße)[404]

In the presence of the Propaganda Minister of the Reich, Goebbels, and the Mayor of Frankfurt, Krebs, Hitler gave a speech to spur the workers on and to justify the project. In addition to the short and middle-term objectives such as decreasing unemployment, he also spoke of the visionary character of the expressway, and that its significance for the German traffic system and economy would only be fully recognized and appreciated in later decades. *„In Jahrzehnten wird man dann den Verkehr abhängig sehen von diesen neuen großen Verkehrsstraßen"*("In a few decades we will see traffic depend upon these large thoroughfares.")[405] The start of constructions for the route Frankfurt–Darmstadt with Hitler's

403 Schütz and Gruber, op. cit., pp. 40–43.

404 ISG.

405 Domarus, op. cit., p.302.

groundbreaking, was exploited on a grand scale by the media-based propaganda; the day concluded with a mass rally in the NS-district Hessen-Nassau in Frankfurt a. M. with a purported 200 000 participants. Goebbels reported about cheering crowds.[406]

The question about the extent to which expressway construction activities could reduce unemployment had already been discussed by the Hafraba. For the construction of the Frankfurt a. M.–Darmstadt segment, the goal was to employ as many unskilled job seekers as possible. Heavy equipment was dispensed with to drive this number higher. This policy had been spelled out on 29 November 1933 by the *Reichsautobahndirektion* (Reich Expressway Directorate) in its note to the *Obersten Bauleitungen.* In fact, the deployment of excavators, for especially difficult terrain, required Todt's approval.[407] Road surfacing – the fourth construction phase – demanded qualified workers who could operate the required equipment. In his research on the participation of the construction company Strabag AG in the building of the expressways, Manfred Pohl comes to the conclusion that machines were lacking in the concreting stage and that considerable delays ensued. Reports which state that the OBK Frankfurt a. M. had threatened, because of the delays, to have the chairmen of the Strabag sent to concentration camps, seem rather unlikely according to records from Hessia's State Archives.[408] No such threats are recorded.

The first stage of the construction works began with the clearing of the woodlands. Tree stumps and roots were extracted from the soil using pulleys and jack stands, and partly, with the help of steam locomotives. A 40 meter wide and 15 kilometers long trail was hewn from the city forest. This work was completed by the end of November. The clearing of the road-way continued until Darmstadt was reached.

406 E. Schütz and E. Gruber: in the place cited, pp. 41–43. See Goebbel's diary entry of 24 September1933 and 25 September 1933, in: Elke Fröhlich (ed.): Die Tagebücher von Josef Goebbels, volume 2, Part 1, München 1987, p. 464s. Goebbels had the photograph of the first cut of spade printed as a postcard and distributed for the Winterhilfswerk (Winter Welfare).

407 HHStAW, Abt. 485, Nr. 246.

408 Manfred POHL, Die Strabag 1923 bis 1998, München 1998, here p. 156.

Image 31: Earthworks on the route Frankfurt–Darmstadt in Winter 1933 (propaganda photograph)[409]

In a second stage, the bridge constructions that were necessary for an intersection-free expressway, were put into place. Fifty-four bridges were needed up to Darmstadt. And fifty-four construction companies using skilled workers, were involved in the completion. The first major construction work was the bridging of the Schwanheimer Straße with the ramp for the Mainbrücke (Main Bridge). The crossing of Goldsteiner Weg was followed by an under-cutting of the railroad line Frankfurt a. M.–Mainz and the large road bridge under the Chausee Frankfurt–Groß Gerau.

After construction of the necessary water ducts, in a third stage, a host of formerly unemployed unskilled workers began with the actual earthworks by excavating clefts into ground waves, thereby obtaining the earth needed for the dam fillings. At times, more than 1600 workers were employed in this section. The reporter Leitner emphasized that on this route not a single excavator was being used. Beginning of November 1933 the construction of the route Mannheim–Heidelberg was begun in four lots. Swastikas hung over intersection constructions and over the barracks of the earth-working gangs.[410] The construction of both

409 Leitner: Reichsautobahnen im Bau, Main-Neckar, Die Autobahn 1934, issue 2, p. 52.

410 Leitner: Reichsautobahnen im Bau, Main-Neckar, Die Autobahn 1933, issue 11, p.14.

bridges over the rivers Main and Neckar began in April 1934.[411] Large-scale propaganda efforts continued to follow the work in progress. Todt inspected the expressway construction sites from the airship "Graf Zeppelin" and let this be reported in the press.[412]

To enable picnics on the grass shoulders of the expressway (see Image 22), Todt instructed the *Gesellschaft Reichsautobahnen* on 24 May 1934 to widen – wherever economically sensible – these shoulders to permit automobile parking. These should be provided particularly at lookouts, lakes, and woods were motorists would enjoy stopping for a break, cf. Image 22.[413]

Todt would further intensify propaganda efforts during the construction of the Munich–Salzburg route by incorporating art and culture into the project. He strongly encouraged "*Autobahnmalerei"* (expressway painting – see section 21), but these aspects are much less visible in the sections of the Rhine-Main-Neckar area. Apart from an exhibition in the *Mannheimer Schloßmuseum*, it was merely the existence of a training and working camp of the *Deutschen Studentenbundes* (German Student Alliance) – connected to Städelschule in Frankfurt – at the construction site of the Urselbachtal Bridge which points in this direction The camp fell under the direction of the Gau training delegates in June 1936. In addition to sports and ideological indoctrination, the "permanent view of the construction work shaped through the power of the Third Reich had an educational effect on the art students. Many hours were dedicated every day to painted and drawn studies of the expressway". The finished pictures were exhibited at Thanksgiving on 4 October 1936 in Niederursel.[414] The Ursertalbach Bridge was erected out of steel, despite material shortages due to the military build-up. The same bridge was blasted by German troops in the last days of war.[415]

---

411 Hans Lindner's report for the magazine "Die Autobahn" describes the state of the construction works on these two bridges beginning of July 1934. Die Autobahn 1934, issue 12, pp. 54–543. About the construction of the Neckar Bridge see also Hanspeter Rings*:* Die Reichautobahn – Monument der Stärke? In: Monika Ryll: Architektur in Mannheim, Mannheim 1994, p. 204.

412 Der Führer, 22 May 1934, Stadtarchiv Karlsruhe, 1/H-1596. Todt reported from the airship "of a splendid view of sweeping lines – an image that appears much more imposing than from the earth." This press report was apparently published throughout the Reich, since it also appears in the press of Munich, see Münchener Zeitung on 22 May 1934, Stadtarchiv München, Presseausschnittsammlung 1933–1934.

413 HHSAW 10.

414 NS-Briefe – Schulungsblätter der NSDAP im Rhein-Main-Gebiet, December 1936, p. 446.

415 Brücke und Strasse, issue 11, 1950, p. 237.

Image 32: Construction site of the Urselbachtal Bridge[416]

[416] Sketch made by the members of the Student's Alliance, Post and Hill.

15 The Main Construction Directorate in Frankfurt

The construction of expressways in the Rhine-Main area was coordinated by the „*Oberste Bauleitung für Kraftwagenbahnen Frankfurt–Heidelberg/Mannheim*"(OBK Frankfurt) (Main Directorate of Expressways Constructions). Its offices were located at the Hohenzollern-Anlage 35 (today: Friedrich-Ebert-Anlage) in Frankfurt a. M. The OBK was founded on 24 June 1933 by a decision of Dorpmüller. On the same date the *Reichsbahnoberrat* (leading official of the *Reichbahn*) Alfred Pückel from Wuppertal was appointed head of the OBK Frankfurt.[417] Kado, before *Oberbaudezernent* (Head of Construction Department) and director at the *Reichsbahn* directorate Frankfurt am Main, succeeded Pückel on 25 May 1937.

The implementation of the specific construction plans at the level of the contract sections was very much shaped by NS-ideology. The *Reichsautobahndirektion* in Berlin demanded on 22 December 1933, that only a nationalist-minded work force be employed.[418] The dog fighting among construction companies for assignments went as far as accusing one another of being Jewish companies or of being financed by foreign capital. In order to return to an orderly course of events, the Reich's government was repeatedly obliged to point out that the Aryan legislation only applied to government organizations but not to private enterprises.[419] Hence the *Reichswirtschaftsminister* (Reich Minister of the Economy) distributed a press release on 15 September 1933 – eight days before the official start of constructions – in which he indicated that the "provision of employment was more important than the Aryan certificate" and explained that the issuance of "official certifications" to companies would lead to misuse and that the subordinate authorities had decided against it.[420]

The shortage of labor which had been becoming evident since 1936 due to the Four Year Plan for Rearmament, also led to a demand for construction technicians at the sections Würzburg,

---

417 Frankfurter Zeitung on 25 June 1933, Stadtarchiv Karlsruhe, 1/H-1596, Der Aufbau der Gesellschaft Reichsautobahnen, Berlin 1935, p. 43.

418 HHSAW 10.

419 Directive of the Reich Interior Minister from 17 January 1934 to the supreme authorities of the Reich, Reich governors, Reich governments and Ministers, which the Reichsbahn passed on to its directorates by way of letter on 13 February 1934. (HHSAW 10).

420 Frankfurter Nachrichten on 15 September 1933, ISG.

Diez and Limburg sites that could not be met. At a meeting of the department heads and the chairmen of the construction departments of the OBK on 23 November 1937 in Frankfurt a. M., it was resolved to increase recruiting efforts for students at technical universities and also to employ candidates with merely adequate grades.[421] In a letter to Kado, director of constructions for the OBK-Frankfurt, Todt expressed his concern that the expressway construction on the route Frankfurt a. M.–Cologne could be halted because of labor shortages and asked Kado to assign the remaining construction lots as fast as possible.[422] When, in 1938, workers were drawn away for fortification activities at the French border (directed by Todt from his office in Wiesbaden) and also for the later defense constructions of the *Organisation Todt* (OT),[423] the shortages intensified, and the construction works had to be slowed down. The well-rehearsed management of work on the expressway's construction sites could be geared towards the defense engineering targets of the OT gradually and seamlessly.

After the completion – or rather freeze – of constructions on the expressway and the dissolution of its construction department on 1 May 1942, the only task remaining for the head of construction management of the *Reichsautobahn Frankfurt* was to ensure the operation and safety of the expressway.[424] For the workers of the *Organisation Todt*, which had been removed from their field, it still provided a link to their "home". It also published the "*OBK-Feldpostbriefe"* (OBK-Troop Letter). Image 33 Shows a cover from the March/April 1943 edition with the combination of low-technology peasantry, modern industry and war machinery that was typical of the Nazi era.[425]

On 16 March 1942, the head of the Construction Directorate of Frankfurt, *Oberbaudirektor* Kado, gave a lecture in the Städelschule Frankfurt a. M.[426] before the members of the *Fachgruppe Bauwesen* (Technical Unit Civil Engineering) and the *Reichskammer der Bildende Künste* (Reich Chamber for the Fine Arts), about the construction of the expressways, using

---

421 HHSAW 10.

422 Ibidem.

423 BA, NS26/1190.

424 Klaus-Peter Barth and Michael Antenbrink: 50 Jahre Frankfurter Kreuz – Von der Nur-Autostrasse zum europäischen Verkehrsknotenpunkt, in: Strassen-Verkehrstechnik, issue 6, 2006, pp. 341–348.

425 HHSAW 10.

426 Ibidem.

slides. On the occasion of the Christmas celebration of the OBK Frankfurt on 18 December 1943, which took place in the hall of the former lodge "*Einigkeit*" ("Unity") at Kaiserstreet 37 in Frankfurt a. M., the OBK invited the family members of the Wehrmacht and OT comrades.[427]

Image 33: Military Letter of the OBK

[427] Ibidem.

During the war years, the traffic on the expressways had been diminishing continuously, forcing the OBK-Frankfurt to shut down one gasoline station after the other. In May 1940, the auxiliary gas station Giessen was closed due to the drafting of a great part of its attendants. The last to be closed was the Lorsch service station in September 1943.[428] In complete disregard for its intended purpose, it was now being used by pedestrians, bicycle riders, horse and cow carts operators , as the OBK-Frankfurt remarked sourly in a letter to the *Landratsamt* (District Office) Groß Geraus on 20 April 1944.[429] In July 1944, to draft additional soldiers for the war mission, the *Reichsautobahn* company was almost completely dissolved. [430] The company's head office in Berlin, however, led a shadow existence until the spring of1945, as recorded by file R4602 in the *Bundesarchiv*.

428 HHSAW 441.

429 Hessisches Staatsarchiv Darmstadt, G15 Kreisamt Groß Gerau, X113, Bau der Autobahn Strecke Frankfurt–Darmstadt 1933–1934.

430 Dissolution of the company Reichsautobahnen due to a decree about total warfare from the "Führer" on 25 July 1944, in: BA R2/25018.

16 The Deployment of Workforce

The question about the extent to which the construction of the expressway could serve to provide work for the unemployed had already been discussed by the Hafraba. For the construction of the route Frankfurt a. M.–Darmstadt, the goal had been to employ as many unskilled job seekers as possible. For this reason excavators and other large machines were largely done without. Until July 1934, the wages for the construction workers on the expressway sites were subsidized – as public relief works – with funds from unemployment insurance – the expressway project thus was co-funded by the unemployment office.[431] Public relief works allowed for the compulsory service of the unemployed through the corresponding employment bureaus. The public relief workers were paid according to the low social-benefit rates, and deductions for food and accommodations in primitive barracks were made from their wages. They were often worse off than social welfare recipients because the latter received additional rent allowances. In the press, accusations were sporadically made that "National Socialism ... erects its constructions through slave work".[432] To counteract this, in February 1934 on the route Echterdingen-Elchingen in Württemberg, a *Streckentarif* – or wage rate, was fixed for construction workers and provided an hourly pay of 60 to 86 Pfennigs. This was higher than the basic wages the *Reichstarifvertrags* (Reich Wage Agreements) had set for building, concrete, and underground engineering construction workers.[433] According to sources, however, the *Streckentarif* remained an odd exception and wasn't applied to the section Frankfurt a. M.–Mannheim–Heidelberg.

The martial language with which the press reported on the beginning and development of the project unwittingly threw a harsh light on the bad working conditions at the construction sites. March 21st of 1934 marked the beginning of the construction activities and was designated "a great action day of the working battle." Terms such as "fight", "gigantic achievement" or

---

[431] Decree from Todt on 20 October 1933, in: BA, R4601/5001. To the following see also E. Schütz and E. Gruber: Mythos Reichsautobahn, Berlin 1996, 2. edition Berlin 2000, chapter „Sozialismus des Strassenbaus".

[432] Der Führer, 16 February 1934, Stadtarchiv Karlsruhe, 1/H-1596.

[433] Ibidem.

"gigantic working battle" were used.[434] In this respect the expressway project even seemed to anticipate the war. In a marked contrast to the glitzy world of expressway propaganda, the morale of the workers was quite miserable.

The worker's situation improved only slightly on 1 July 1934, when the employment contracts were transformed into free contracts and the *Reichstarifvertrag* was applied. This was owing to the enterprises' demand for qualified construction workers who could more easily withstand the pressures of the hard labor. Conflicts over this arose again and again because employment bureaus were also procuring fewer capable workers who could tolerate the hard working conditions on the construction sites without falling ill frequently.[435] The coeval *Deutschlandberichte* (Germany Reports) of the Social Democratic Party in 1935 reported on the drudgery, and an appalling booster system together with the miserable conditions for wages, food and accommodations at a site in Mannheim.[436]

The workers were often separated from their families by hundreds of kilometers and lived in primitive barrack-camps at the expressway sites. As a result of the labor shortage the construction workers' camps were improved to provide better lodging. The extensive media coverage at this time about the allegedly lavishly furnished camps was striking. To soften the plight of the camp dwellers, the organization *Kraft durch Freude* provided the expressway working camps in Northern Hesse with an entertainment program. Three times a month a theater group livened up each of the camps.[437] Hohlmann has identified six former expressway camps in the area of Bad Hersfeld alone, in which forced laborers of the armament industry were lodged.[438]

The President of the *Reichsanstalt für Arbeitsvermittlung und Arbeitslosenversicherung* (Reich Institute for Job Placement and Unemployment Insurance) in his letter to the President

---

434 Badische Presse on 21 March 1934, Stadtarchiv Karlsruhe, 1/H-1596.

435 Announcement of the Inspector General for Road and Traffic from3 July 1934 (HHSAW 10).

436 Citation from *Rings,* op. cit., p. 206s.

437 Kraft durch Freude, Gau Kurhessen, February 1937 program, p. 39.

438 Susanne Hohlmann: Pfaffenwald – Sterbe- und Geburtenlager 1942–1945, Nationalsozialismus in Nordhessen vol. 2, Kassel 1983, p. 30: On the camps in the area of the Main Direction of Constructions in Kassel see also the interview with their superintendent, Liebetreu, in the Kasseler Post of 19 June 1937. Liebetrau names 11 camp sites.

of the *Landesarbeitsämter (*State Employment Bureaus) on 2 January 1935, deplored that the allotted workers often lacked due willingness to work and that this would inevitably lead to dismissals.[439] He also stated that the enlisted workers had not been well-enough informed about the poor wages and housing conditions at the expressways. On 11 January 1935, the OBK-Frankfurt responded that their district did not house workers in barracks because they engaged men who lived in close proximity to the work site. It also lamented that the health insurance companies that covered workers from the construction sites were very troubled by the sick reports. The workers with stomach ailments, for example, were not able to receive a special diet from the common kitchens. The president warned of the recruitment of politically unreliable workers from the major cities and referred to work stoppages in Camp Gyhum on the Hamburg–Bremen section, where unemployed Berliners had been deployed.

The *Landesgewerbearzt* (Physician of the Trade Supervisory Center) of Baden reported on 13 August 1935 with concern about an ailment that afflicts trench diggers and had befallen expressway workers. This situation was also reported by the journal *Die Straße.* "As a result of exhaustive digging, a tearing of the spinal prolongation of the lower cervical and upper thoracic spine may occur. It is caused by muscle contraction from strenuous digging. Disability lasts six weeks or more and has a negative economic impact. Approx. 10% of the work force is afflicted by this ailment.[440] To prevent an exodus of even worse paid rural workers from the farms, the *Reichsautobahndirektion* sent a circular on 10 March 1934 to all *Obersten Bauleitungen* with instructions that no unskilled rural laborers be deployed for the construction of the expressways.[441]

The xenophobic ideology of the NS-regime prevented the effective integration of labor gangs from other parts of the country. Hence the *Kreisleitung* (district committee) Darmstadt of the NS-*Gauleitung* Hessen-Nassau complained to the OBK-Frankfurt on 6 April 1935, that Bavarian companies were employed in the area. It lamented that the local workers were hardly able to understand the dialect and manner of the site manager and foreman from

439 HHSAW 259.
440 HHSAW 259.
441 Ibidem.

Bavaria.[442] For 18 December 1936, Todt arranged a joint Christmas celebration for all expressway workers employed by the various construction companies. He announced his intention to speak to the celebrating workforce between 18.50 and 19.00 via the Reich radio broadcasting system.[443]

After the rearmament program had already led to a labor shortage, the provision for manual labor for earthworks was renounced. Heavy machinery now made work easier. Noteworthy are the many references, in reports about the construction of the expressway in Hessen for example, of multiple production shifts on various sections and constructions – in the winter these were carried out with artificial lighting.[444] The urgency of the entire project derived from the propaganda target of completing 1000 km of route a year beginning in 1936. Bridge constructions such as the expressway bridge over the Neckar near Mannheim, were projects that spanned several years and had to be built in multiple production shifts.[445] Sections outside of Hessen also made use of several work shifts – sometimes even three-shift-operations.[446]

In addition to the regular forced laborers who were engaged in expressway construction, there were also workers from the occupied territories who had let themselves be recruited in good faith at the very beginning of the war as "foreign workers." This can be verified in the expressway camp Pflanzgarten near Viernheim, for example.[447] Because Todt had classified the expressway Mannheim/Saarbrücken as essential to the war effort on 20 December 1939, construction work there was able to continue, but the labor shortage led to the need for additional workers from the occupied territories – in NS-terminology – *Fremdarbeiter*

---

442 Ibidem.

443 HHSAW 10.

444 Bridge over the Theistal near Niederhausen as per Höchster Kreisblatt on 22 July 1939 (AHR), section Frankfurt–Bad Nauheim as per General Anzeigerfrom 26 August 1936 (AHR), Badischer Beobachter on 21 November 1933, Stadtarchiv Karlsruhe, 1/H-1596, press release of Todt's on 3 October 1935 about opening of the route Frankfurt–Mannheim–Heidelberg, Neueste Zeitung on 3 October 1935 (AHR), construction of the Asbachtalbrücke near Bad Hersfeld as per Hohlmann, op. cit., p. 42.

445 „Die Reichsautobahn in Baden", in: Heidelberger Neueste Nachrichten on 1 August 1935, Stadtarchiv Heidelberg, Zeitungsausschnittsammlung zum Autobahnbau 1933–1938.

446 Friedrich Doll: Vom Bau der Reichsautobahn München–Landesgrenze, in: Die Strasse, issue 5, 1935, p. 157.

447 Hessisches Staatsarchiv Darmstadt, Akte R21L, Quellen zur Zwangsarbeit, G15, Heppenheim G328. To the deployment of forced laborers see Ulrich Herbert: Fremdarbeiter – Politik und Praxis des "Ausländer-Einsatzes" in der Kriegswirtschaft des Dritten Reiches, Bonn 1999.

(foreign workers). The expressway construction section from Viernheim to Frankenthal was executed by the building contractors Kallenbach and Grün, which maintained a working camp called Pflanzgarten near Viernheim. Here lived 250 workers from Poland and Czechoslovakia who had been on their way to the construction section Limburg in winter 1939, when they were transferred to Viernheim according to the instructions of the OBK Frankfurt. The arrival of these workers, who had allegedly registered voluntarily for service, brought about considerable confusion in the xenophobic German Reich. They were seen virtually as enemies but they couldn't be treated as prisoners of war based on the *Ausländergesetz* (Law on Foreigners) from 1 October 1933. This was ascertained in writing by Mannheim's employment bureau from the *Landratsamt* (district office) Bergstrasse on 24 February 1940. However their freedom of movement was severely restricted which can partly be explained by the existence of very important military facilities, including munitions dumps, near Viernheim. The worker's freedom of movement was restricted and they were not permitted to leave the site of Viernheim in their free time or, above all, to go to Mannheim. They were also not allowed to frequent public houses after 8 p.m. To visit cinemas and dancing halls was completely forbidden. The transport of laborers from the occupied territories to Limburg also indicates that the construction of the expressway was being continued with "foreign workers" there as well.

17 The Frankfurt Intersection

The *Frankfurter Kreuz* (Frankfurt Intersection) with its cloverleaf-shaped traffic routing has become a symbol for modern expressway traffic systems. It was, however, neither the first completed expressway intersection in Germany nor was it the first with a cloverleaf shape. The cloverleaf-shaped Schkeuditzer Kreuz near Leipzig, which was opened to traffic in April 1936, preceeded it.[448]

Image 34: Sparse Traffic at the Frankfurt Intersection on 2nd April 1957[449]

The planning of the *Frankfurter Kreuz* was from the beginning closely linked to the re-settlement of the airport of Frankfurt. During the 1920s, Frankfurt's airport in Rebstock had reached its capacity limits. The Chamber of Commerce and Industry Frankfurt a. M.-Hanau was not particularly active in the discussion about the extension of this airport, if the reports

[448] Die Strasse, issue 8, 1936, pp. 228–242.

[449] Darchinger 2008, p. 169, with permission of the Friedrich-Ebert-Stiftung.

published monthly are assessed.[450] The Chamber of Commerce and Industry of Düsseldorf, however, represented many companies from the industries of consumer products and finished goods, which, due to their export orientation, had developed a vital interest in the expansion of air traffic. Accordingly, there were conspicuously many publications in the Düsseldorf chamber magazine *Wirtschaft und Verkehr* (Economy and Transport) about the significance of air transport.[451] It is therefore not surprising that for the Reich-wide planning of German airport capacities, Düsseldorf seemed to have an edge over Frankfurt. However, Gauleiter Sprenger could, pointing to the prominent industrial location of chemical trust "IG Farben" in Frankfurt, push through the build-out of the airport of Frankfurt, that was relocated in 1936 to the city woods near the planned Frankfurt expressway intersection, and on 8 July 1938 was opened as "*Flug- und Luftschiffhafen Rhein-Main*" (Flight and Airship Harbor Rhine-Main)[452]

In NS letters-of-indoctrination and in a newspaper article, Director Pückel from the *Obersten Bauleitung* in Frankfurt a. M. introduced the connection *Frankfurter Kreuz* and Airship Harbor Rhine-Main ("world airport") and hailed it the most modern of traffic junctions.[453] On 11 May 1936 the first zeppelin landed at the new airport and drew a great crowd of onlookers.[454] The expressway turnouts near Frankfurt Airport were congested. On 10 May May 1936, at a meeting in the building of the *Reichsbahndirektion* Frankfurt a. M. of the OBK with the Hessian state government, police and airport management, this problem was discussed and the proposal was made to build an access to the airport at kilometer 1.8.[455]

The *Frankfurter Kreuz* was indeed conceived at the beginning of the NS expressway project, but the route Erfurt–Saarbrücken was planned to be the east-west line.[456] At a meeting of the *Provinzialdirektion Rheinhessen* (Head of Provinces Rhinehessen) in Mainz on 17 October

---

450 Hessisches Wirtschaftsarchiv Darmstadt.

451 Ibidem.

452 Michael Wustrack: Vom Flughof Rebstock zur Drehscheibe im Weltluftverkehr, in: Ulrich Eisenbach and Gerd Hardach (eds.): Reisebilder aus Hessen, Fremdenverkehr, Tourismus und Kur seit dem 18. Jahrhundert, Darmstadt 2001, pp. 163–176, and Zibell, see: Jakob Sprenger (1884–1945) – NS-Gauleiter und Reichsstatthalter in Hessen, Darmstadt 1999, p. 312.

453 Generalanzeiger on 17 December 1936 (AHR), NS-Briefe – Schulungsblätter der NSDAP im Rhein-Main-Gebiet, Dezember 1936, p. 441s.

454 Ibidem, p. 447.

455 HHSAW 6.

456 Frankfurter Nachrichten on 25 September 1933, ISG.

1933, the intention was articulated that the east-west route should cross the north-south route at Rüsselsheim. The management of the Opel car plants demanded that the route be laid as close as possible to the town Rüsselsheim.[457]

According to research done by Barth and Antenbrink, in 1934 already the route Würzburg–Frankfurt–Cologne had been established for the east-west connection.[458] In the years 1934 and 1935 a discussion unfolded about the design of the intersection: as a double trumpet or as a cloverleaf. The latter was sanctioned in early January 1936 by the *Reichsautobahn* administration. A problem emerged with the junction "Frankfurt Süd" of the route Frankfurt–Darmstadt, which already in 1935, when the route was opened, had been laid out at the Reichstrasse 43. The junction lay within the range of the cloverleaf and had to be forfeited with construction of the intersection, which was not started until the early 1950s. Image 35 shows a photograph of the junction Frankfurt Süd from 1935.[459]

Image 35: Junction Frankfurt Süd in 1935

457 Hessisches Staatsarchiv Darmstadt, G15 Kreisamt Groß Gerau, X113, Bau der Autobahn Strecke Frankfurt–Darmstadt 1933–1934.

458 Klaus-Peter Barth and Michael Antenbrink: 50 Jahre Frankfurter Kreuz – Von der Nur-Autostrasse zum europäischen Verkehrsknotenpunkt, in: Strassen-Verkehrstechnik, issue 6, 2006, pp. 341–348.

459 Klaus-Peter Barth and Michael Antenbrink, in the place cited.

From the east-west line Aschaffenburg–Frankfurt–Köln, only the section Frankfurt–Limburg a. d. Lahn was completed by 1939. It could be reached from Frankfurt via an access to the beltway Frankfurt–Wiesbaden north of the Main. Despite the German attack on Poland on 1 September 1939, the expressway project was continued in an attempt to maintain a semblance of normality. The route Frankfurt–Limburg was opened symbolically on the Sixth anniversary of Hitler's first cut of spade, on 23 September 1939, in times of "high political tension" as the *Frankfurter Wochenschau* wrote.[460] In Limburg the route ended – for the time being – with a 500 meter long expressway bridge over the Lahn, which was made of 13 brick arches of 60 m height, and ranked as the longest expressway bridge made of stone in Germany.[461] Paul Bonatz had designed the bridge and had placed it as a "counter-image" to the Limburg cathedral, which until then had dominated the view over the valley of the Lahn. Image 36 shows the bridge with sacral-looking, soaring pillars opposite the cathedral in a photograph from 1940.[462] When the German troops retreated, the bridge was blasted in the last days of war. Image 37 shows a photograph from the year 1945.[463] The entire route Frankfurt–Cologne was opened on 15 June 1940.[464] The routing across the bridge had the "price" of a steep ascent on the Elzer hill – proving again that aspects of landscape aesthetics took priority over transport engineering.[465]

---

460 Frankfurter Wochenschau Nr. 39, 1939, ISG, S3/7461.

461 To the bridge see the study of Jennifer Verhoeven: „...hier war einmal Vollkommenheit erreicht." Die Reichsautobahnbrücke von Paul Bonatz bei Limburg a.d. Lahn, in: Denkmalpflege und Kulturgeschichte, issue 2, 2007, pp. 2– 8 and Kuhlmann 2007.

462 From: Paul Bonatz zum Gedenken, Technische Hochschule Stuttgart, 1957.

463 Heinz Maibach: Dokumente zur Limburger Stadt- und Kreisgeschichte, Stadtarchiv Limburg 1992, p. 160.

464 Schwell and Jäger, p. 481.

465 Kuhlmann 2007, p. 148.

Image 36: Lahn Bridge near Limburg 1940

The construction of the *Frankfurter Autobahn Kreuz* became tangible for the first time in 1938, when the route Frankfurt a. M.–Aschaffenburg was first tackled. To this avail, three new working camps were set up on the route with the names Kelsterwald, Mönchwald and Steingrund, as described by the *Frankfurter General Anzeiger* in a report.[466] The construction works on the east-west route south of the Main began in April 1939. On 22 November 1939 the *Reichsautobahn* administration authorized the implementation planning of the *Frankfurter Kreuz*, which envisaged an eight meter wide ramp.

466 Frankfurter General Anzeiger on 18 August 1939, AHR

Image 37: Destroyed Lahn Bridge 1945

The *Frankfurter Kreuz* had just reached the stage of earthworks when, on 20 October 1941, the construction works were brought to a halt due to the war.[467] It was not until 1956 that the intersection of east-west and north-south bound expressways near Frankfurt was realized by the Federal Republic of Germany: On 10 July 1956, with Willy Hof as the guest of honor, the Federal Transport Minister Seebohn inaugurated the intersection.[468] In the years before the route was opened, there had been frequent initiatives and proposals to complete the missing section of the route Cologne–Frankfurt from the Frankfurt–Wiesbaden route to the *Frankfurter Kreuz*. Traffic on the expressway Cologne–Frankfurt – with 3000 vehicles a day – was, in fact, still quite low as a traffic census in 1950 revealed and the Image 34 shows. However,

---

467 Barth and Antenbrink.

468 Deutsche Messezeitung, Nr. 7, 1956, ISG, S3/24863. Hof was Director-General of the Deutsche Handelsgesellschaft in Frankfurt, as well as a racing driver, and had been managing director of the Hafraba, see biographical notes on Hof's 70th birthday on 15 November 1950, in: Brücke und Strasse, issue 11, 1950, p. 226.

there was much pressure to "close the gap". Because of the federal government's lack of funds, the construction took several years.[469]

In the *Frankfurter Kreuz*, the lavishness and extravagance of traffic planning and implementation in the NS-era becomes apparent; it developed apparently independently from economic and financial bottlenecks which resulted from armament policies. City councilman Schlötter's prediction that Frankfurt a. M. would become a “city of roads” would prove to be true. Not only the Main Bridge was built within the section *Frankfurter Kreuz* to Bad Nauheim; also, north of the Main, the beltway Frankfurt a. M.–Wiesbaden and a further bridge over the Main near Eddersheim were built as part of the *Autobahn* 3 from Frankfurt to Limburg. The construction of this bridge was begun on December 1938. In August 1939 the river pylons were to be stonewalled with massive blocks of stone which had been extracted from sandstone quarries in Edenkoben in southern Palatine. But the substructure and construction of the pylons were protracted until 1943, when the construction had to be completely halted due to the war.[470]

The fact that the planned route from *Frankfurter Kreuz* to Limburg would make a further bridge over the Main and a 20 km long section to the beltway Frankfurt a. M.–Wiesbaden necessary, illustrates the waste of resources which resulted from the planning. It would have been much less costly to build an expressway routing connecting the route from Frankfurt a. M. to Limburg to the route Frankfurt–Bad Nauheim over the beltway Frankfurt a. M.–Wiesbaden north of the Main (see Image 38). It is intriguing that the Federal Republic of Germany implemented the "prestige project" of a separate stretch to join the route to Limburg, as well as the second bridge over the Main, as planned in the NS-era, in the early 1950s when financial resources of the federal government were acutely limited and traffic on the expressway was relatively low. This can count as evidence for the continuity of large-scale projects across deep historical caesura.

---

469 Inquiry of the Mannheim's representative Margulies on 13 October 1950 about the closing of the gap at the Frankfurter Kreuz to the Federal Transport Minister, in: Brücke und Strasse, issue 12, 1950, p. 247. To the traffic count ibidem, issue 7, 1950, p. 145.

470 Höchster Kreisblatt on 22 July 1939, AHR, HHStAW, Bestand 485, Akte 401.

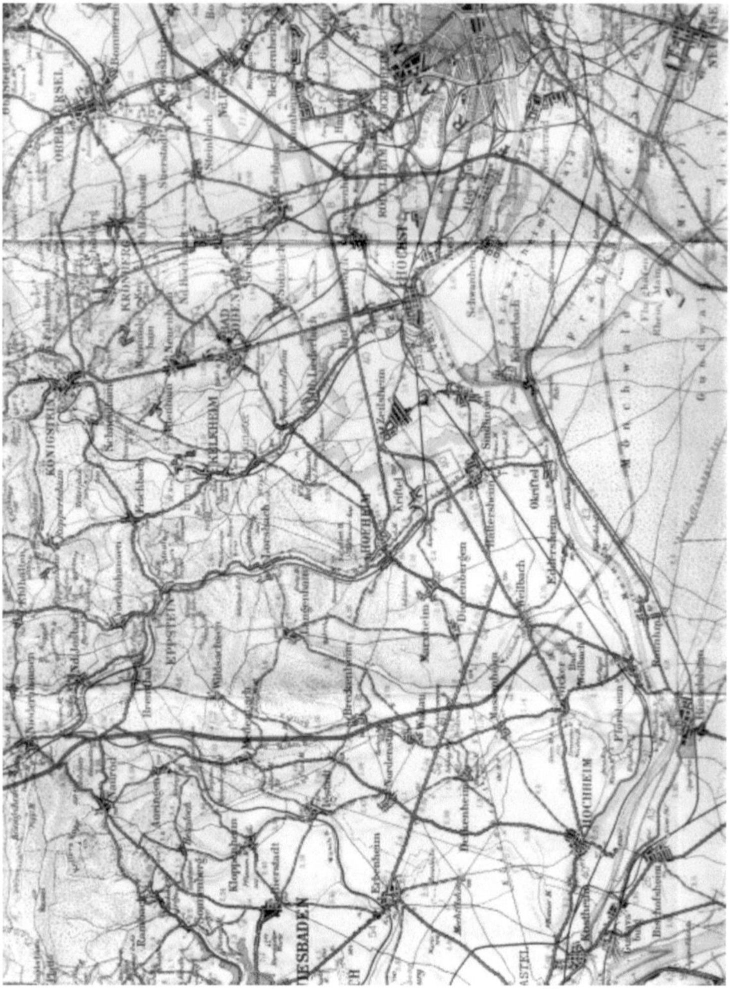

Image 38 Road network in the wider area of Frankfurt in the year 1948[471]

[471] Ravensteins Wander-, Rad- und Autokarte Umgebung von Frankfurt am Main, 1948.

18 The Route Darmstadt–Heidelberg–Karlsruhe

On 3 October 1935, Todt opened the route Darmstadt–Mannheim–Heidelberg after he had already made a preliminary viewing, together with a large retinue, on 23 September – the first anniversary of the ground-breaking. The Gauleiter, the mayor, representatives of the *Reichsbahn* and Hof, as the representative of the *Reichsautobahngesellschaft* (Reich Expressway Company), as well as delegations from Goebbel's ministry and the *Wehrmacht* took part. Philipp Prinz from Hesse was dispatched as head of the *Motorbrigade* Hessen and representative of the NSKK-*Korpsführer (*NSSK-Corpsleader*)* Hühnlein. A column of 50 "sleek automobiles" with Todt's "well-known large car" at the head, started their ride to Mannheim and Heidelberg at kilometer zero.[472]

The Frankfurt–Heidelberg route was used for various automobile races. In the fall of 1935, Ernst Henne made a first attempt to break a speed record on this stretch of the Reich expressway. In October 1937, speed record holders from all over Europe were invited for the International Week of Records for Automobiles and Motorcycles. Hans Stuck attempted a speed record on the Frankfurt–Heidelberg route on 23 March 1936 and reached a top speed of 312 kilometers per hour.[473]

The twin cities Mannheim and Heidelberg were now joined to the expressway, as had been aspired to since 1926. An exhibition in the *Mannheimer Schloßmuseum* with the title "From Wild Path to Reich Expressways" presented archaeological finds from the construction works.[474] Mannheim's pride at having a connection to the expressway was revealed by the embellishnents that it provided out of city funds, cf. Image 39. The entrance area was furnished with two high pylons beneath bronze Reich eagles with wreaths and golden swastikas and the lettering "*Reichsautobahn*". In a strange convergence of modern transport

---

[472] Neue Mannheimer Zeitung on 24 September 1935, Stadtarchiv Mannheim, Zeitungsausschnitte S2/1524.

[473] Reuss, p.189.

[474] For the following part see: Hanspeter Rings*:* Die Reichautobahn – Monument der Stärke?, in: Monika Ryll*:* Architektur in Mannheim, Mannheim 1994, pp. 202–210.

technology and a backward political institution, Mannheim's NS newspaper, "*Hakenkreuz-banner*", called the entrance "royal".[475]

Image 39: Entry to the expressway in Mannheim 1935[476]

The plans intended to develop Mannheim as an expressway intersection to Saarbrücken. As late as 1940 during the ongoing war, and in spite of concomitant material shortages the steel bridge over the Rhine near Sandhofen was being built; it would, however, collapse due to construction faults.[477]

In 1938 the journal *Die Strasse* conducted a survey among leading officials of the NS-regime about the significance of the expressway. Jakob Sprenger, district head of Hesse-Nassau, explained: „*The Reichsautobahn* is a means of promoting the Rhine-Main area. The route Frankfurt a. M.–Darmstadt was the first trial route (...). Experts from all over the world come to inspect it“. He also referred to the fast accessibility of destinations within his district. That was why the expressway was valuable for his party and country.[478]

---

475 Special edition on 3 October 1935 on the opening of the leg Mannheim–Darmstadt, Stadtarchiv Mannheim.

476 Hakenkreuzbanner on 3 October 1935.

477 Ibidem, p.205. See also Brücke und Strasse, issue 10, 1950, p. 214.

478 Die Strasse, issue 15, 1938, p. 475s.

The six expressway routes that were put into operation within the German Reich on 27 September 1936, completed the first 1000 kilometers of the project and fulfilled the predictions of the Propaganda Ministry. Hitler personally celebrated the occasion during a ceremonial act at an expressway location near Breslau. Among the completed routes were the Frankfurt–Bad Nauheim and Heidelberg–Bruchsal legs which now provided an expressway that was negotiable for traffic along approximately160 kilometers. During opening festivities at the gates of Bruchsal, Dr. Freitag, official of the *Reichsbahn*, stressed the importance of this route for tourism to the cities of Heidelberg and Bruchsal. The governing mayors of Heidelberg, Mannheim, Karlsruhe and Bruchsal were among the guests who had assembled for this event.[479] What these inaugural ceremonies had in common was the overt emphasis that was placed on the aspect of construction accomplishments. In this connection, the number of laborers involved and locomotives used was registered along with the excavation volume from the construction sites. Capital expenditure and funding, on the other hand, were not mentioned as this would hardly have offered a suitable projection for propaganda purposes considering the difficulties faced in the borrowing arrangements with the *Reichsbank*.

In its 28 September 1936 edition, the *Völkischer Beobachter* reported on the adaption of this route to the characteristics of this swathe of land with its views of the hilly landscape near Heidelberg, the amphitheater-style Thingstätte and the well-known stone quarries near Dossenheim. Heidelberg itself was joined to this new expressway via the extended old Speyer road which served as a feeder road. An extra effort was made to point out that the old route connecting Mannheim to Karlsruhe was heavily used by 4000 vehicles per day.[480] Sunday traffic saw a near doubling of this number. The considerable time-saving aspect of the route due to its avoidance of travel through the numerous villages was noted. And finally, the newspaper reporter also observed that the vehicle operator would travel on sure ground to his destination.

---

[479] Völkischer Beobachter on 28 September 1936, HHSTAW, 14

[480] According to the Continental-Handbuch für Kraftfahrer, Hannover 1931, p. 972, this is road no. 293 Mannheim–Speyer–Karlsruhe.

Image 40: View trough the Wind Shield of a Bus on the leg Heidelberg–Bruchsal[481]

Another newspaper article on the expressway openings pointed out that expropriations had not been necessary even a single time. Indeed, farmers had even received compensation for a *Flurbereinigung* (or „land consolidation“) which had been carried out. Farmers were also reimbursed where crops were lost.[482] What is remarkable is that a linkage between the construction of the expressways and the *Flurbereinigung* was established right at the beginning of the expressway project. Using this process of land consolidation to facilitate the acquisition of land had been put forth by Willy Hof at a meeting with Hitler as early as 6 April 1933.[483]

On 1 October 1937, Fritz Todt, Inspector General of German Roads, inaugurated the Bruchsal–Karlsruhe leg of the *Reich* expressway. The 230 members of the German Roads Delegation who took part in the celebrations spent eight days touring the German express-ways at the invitation of Todt. With the opening of this new expressway segment, the

---

481 Hakenkreuzbanner on 28 September 1936, HHSTAW, file 485, no. 14, press cuttings.

482 Ibidem.

483 See report from Hafraba e.V.'s managing director, Hof, on his 6 April 1933 meeting with the Reich Chancellor on the planning and financing of the expressway construction, in: Repgen, vol. 1,p. 306s. Die Kasseler Post from 19 August 1934 (Stadtarchiv Kassel) also reported on land compensation for the farmers who were affected.

contiguous route linking Bad Nauheim–Frankfurt–Mannheim–Heidelberg–Karlsruhe reached a length of 175 kilometers.[484]

[484] Die Straße 1937, p.570.

19 Two Concepts for Routing in Hesse

When construction work on the autobahn Frankfurt–Mannheim started in 1933, the detailed plans that had already been drawn up by the Hafraba could be implemented straight away, but it was not possible to incorporate the concepts of landscape aesthetics and monumentality. The autobahn section Frankfurt–Karlsruhe was crossed by functional bridges and runs along long straight sections across flat terrain with only a few curves that have large radii, thus promising safe and comfortable motoring.[485]

Image 41: Functional bridges at the leg Frankfurt–Karlsruhe[486]

Furthermore, towns that are directly accessible from the autobahn are laid out along the route like pearls on a necklace so that motorized local traffic between these towns could be handled over the expressway: Frankfurt–Darmstadt–Mannheim–Heidelberg–Karlsruhe. The *autobahn* passes these towns close to the town boundaries. Following the early Hafraba blueprints for that line had meant that the modern expressway followed the old layout of the railroad dated

485 The routing is described in: R. Auberlen: Die Verbindung des Landstrassennetzes mit der Reichsautobahn Karlsruhe–Frankfurt–Hersfeld, in: Die Strasse, issue 11, 1937, pp.302–307, as well as in issue 9 and 10 (for the inauguration of the section Frankfurt–Darmstadt), 1935.

486 Bauwerke der Reichsautobahn Darmstadt–Mannheim, in: Die Strasse, 1935, p. 748.

from 1850[487]: A direct link between the cities Heidelberg and Mannheim and – in the middle of this link – a junction with the line to Darmstadt, which has been directed in the middle of the land strip of the right side of the Rhine valley. This implied that the densely populated villages at the foot of the hills could not be served by this line.

After 1934 new plans were drawn up for routing the rest of the Hessian network. These were based on the principles of landscape aesthetics and departed completely from the concept that the Frankfurt–Karlsruhe autobahn was based on. For the implementation of the Gezuvor plans, the main directorates were entrusted with the detailed planning of the route and construction. This detailed planning was then carried out in consultation with the *Landräten*, mayors, farmer's organisation, *Landesbauräten* and cultural monument offices,[488] and then submitted for final approval by Todt. In many cases, he made final adjustments to details, as demonstrated by the files of the main construction directorate in Kassel.[489]

The principle of landscape aesthetics realized in the Hessian Network demanded that autobahns, unlike railway lines which cut through hills or cross valleys on raised dams and which therefore stand out from the landscape as technical constructions, should be adapted to the landscape. This kind of routing in accordance with the intention of landscape aesthetics did indeed provide panoramic views of the landscape and of towns – e.g. the approaches to Kassel from slopes and mountain peaks[490] – that had been unknown on railway lines and country roads as these run through valleys whenever possible. In his lecture "From the Hafraba to the *Reichsautobahn*," the section head of the Gezuvor, Becker pointed out that the North Hessian section was deliberately routed over the peaks of the plateau, while valleys were usually crossed vertically. No other routing was possible considering the level of settlement in recent years, he argued.[491] However, this argument is not very convincing if we take into account the lower population density before 1945 compared to the new housing

---

487 Reichsverkehrsministerium (Ed.): Hundert Jahre deutsche Eisenbahnen, second edition, 1938, p. 32.

488 See "Ehe die erste Lore rollt", archive city of Kassel, Kurhessische Landeszeitung on 19 March 1937, archive city of Kassel.

489 in: BA R 4601, file 4043, 4046, 4045.

490 See the reports on the inauguration of the section Göttingen–Kassel: "Zur neuen Schönheit führt ein neuer Weg" in the Kurhessischen Landeszeitung, the newspaper of the Gau Kurhessen, on 19 June 1937, archive city of Kassel.

491 Kasseler Post on 18 March1934 UBK.

projects that have been put up since the 1960s. The routing of sections that pass important tourist attractions such as Neuenstein Castle,[492] the exit Bad Hersfeld-West, and Montabauer Castle on the A3[493] illustrates how roads were fitted into the basic idea of landscape aesthetics.

9. Bauabschnitt Hannover—Kassel. Die fertige Werratalbrücke bei Hegemünden (Hannoverisch-Münden). Mai 1938

Image 42: Bridge over the Werra Valley.[494]

Moreover, one can recognize a remarkable realisation of the principles of landscape aesthetics and monumentality in the way that the *autobahn* exit *Südkreuz* Kassel (south junction), was laid out on a mountain peak that requires an ascent of 80m on both the north and south sides. This can be considered a triumph of Nazi construction of the A7 *autobahn*, which went far

---

492 Neuenstein Castle was mentioned several times in local press reports about the completed Kassel–Bad Hersfeld route, see archive city of Kassel, Kurhessische Landeszeitung, Whit holiday 1938 and on 14 January 1939. The Munich–Salzburg autobahn runs over the bridge over the Mangfall and provides a view of the church at Weyarn with the world famous Annunciation group by Ignaz Günther.

493 I have used the present numbering for the roads. For references to earlier numbering see K. Schefold and A. Neher, op. cit., Anlage 21.

494 Ernst Vollbehr: Arbeitsschlacht – Fünf Jahre Malfahrten auf den Bauplätzen der „Strassen Adolf Hitlers", Berlin 1938.

beyond all conventional road planning visions. The Third Reich's overwhelming desire to assert itself also manifested itself in the gigantic bridges, such as the Werra Valley Bridge near Hannoversch Münden with a steep gradient of 7.8%[495] (cf. Image 42 and Image 74), the Fulda Bridge (only constructed in 1962) with a span of 120m between the piers near Bergshausen on the *Südkreuz* Kassel and the numerous, and extremely expensive large constructions, that support the slope location of the A4 from Bad Hersfeld to Eisenach high over the Werra Valley.[496]

The idea of landscape aesthetics was so highly developed that *autobahn* sections not only hugged valleys and mountains with many narrow curves, they were also laid over countless summits resulting in great panoramic views. The way over summits had gradients of up to 8%.[497] This gradient of 8% is all the more surprising as Todt's decree of 12 December 1933 to the Gezuvor and the *Reichsautobahngesellschaft*, "Guidelines for Autobahn Planning"[498] stated that gradients of 5% or more should be avoided as far as possible. Some months later Todt changed his mind and admitted gradients of more than 5% in routes leading through hills on the grounds that the car passengers could get a lively impression of the beauty of the landscape.[499]

There are many publications on the subject of landscaping the *autobahn* (cf. appendix 2). In theses publication the conflicts among the different professional groups is often stressed: civil engineers who pleaded for cost-effective construction and road safety features versus the architects for landscaping Todt had brought into play. But in the files of the main construction directorate in Kassel in the *Bundesarchiv,* one can not find evidence for the influence of the landscape architects as a professional group. Publications on the subject of landscaping could

---

495 This downhill section is not, however, bought at the price of a view far into the landscape. For more about the bridge at Hannoversch Münden see the Bundesminister für Verkehr (publisher.): Werratalbrücke Hedemünden, Hannover 1993. On the bridges in North Hesse see also the article: "Landschaft gestaltet Brücken" in the Kurhessischen Landeszeitung on 10 September 1935, Stadtarchiv Kassel.

496 The bridge at Wommen was only completed with four lanes in the course of the transport projects after reunification in 1989, see Amt für Straßen- und Verkehrswesen Frankfurt (publisher): Bundesautobahn A4 Bad Hersfeld–Erfurt, Frankfurt 1994.

497 Daten zu den Nordhessischen Autobahnen sind enthalten in Straßenneubauamt Hessen-Nord (Editor): Autobahnbau in Nordhessen, ca. 1974.

498 in: BA R 4601, file 1263, sheet 65–70.

499 Letter of Todt to Gesellschaft Reichsautobahn on 20 February 1934, in: Thüringisches Hauptstaatsarchiv Weimar, C234, p. 24.

be augmented by relating landscaping to road safety which was the priority of the civil engineers. There has been no discussion as of yet on the trade-off between these two goals: Steep gradients significantly reduced the effortless and risk-controlled practicability of the autobahn for automobiles.

I will show the reduction of road safety here with two examples.

(1) The *autobahn*-related term "Kassel Mountains" was coined – these mountains are still feared by motorists today, especially in winter. Truck drivers on the section Göttingen–Bad Hersfeld who wish to avoid the “Kasseler Mountains” still use the B27 along the rivers Werra and Fulda – which raises justified doubts about the sense of the routing in terms of traffic engineering.

(2) The Rimberg[500] service station on the A5 was located at the top of a mountain providing a wonderful panoramic view – but at the price of a steep ascent that represents a grave threat for road safety. The service station is located 500m above the normal sea level at the highest point of the road in Hesse and the ascent is especially hazardous in winter. On the other hand, it would have been easy to bypass the Rimberg by laying this section of the road 500m further to the east, which shows the special orientation of planning to observation points. The frequently cited discourse about landscape aesthetics, however, does not suffice to explain the routing of the autobahn in Hesse. The long-neglected issue of routing must now be addressed. The remarkable aspect about routing in Hesse is that the routing between Frankfurt and Kassel deviates in two respects from the Kassel–Marburg–Giessen–Frankfurt route originally drawn up by the Hafraba, cf. Image 43. Not only was the mountainous route for the A7/A5 chosen – instead of one that would have run through largely flat terrain – but it is also conspicuous that the section that was realized is designed as a **trunk road connection** Kassel–Frankfurt and at the same time rules out local traffic between the medium sized towns. The connection between the two Hessian towns Marburg and Giessen with Kassel provided for in the Hafraba route was dispensed with completely while the two Hessian spa resorts Bad Hersfeld and Bad Nauheim, that were of only minor importance in this period, were connected. Once again, this emphasises the weight attached to the tourist aspects of road

[500] The service station was planned, but only built in 1949, see Bundesminister für Verkehr (publisher): HAFRABA – Bundesautobahnen Hansestädte–Frankfurt–Basel, Bau Verlag, Wiesbaden and Berlin 1962, p. 123.

routing. This routing also makes it clear that autobahn construction did not completely and in every respect serve to meet specific transport needs. If one interprets the routing of the A5/A7 and the A3 from the *Reichsbahn's* perspective, it fits seamlessly into the *Reichsbahn's* interests of not sharing business in either local passenger or freight traffic with rivals from the road transport sector, whereby it has to be remembered that in passenger transport the volume of local traffic far exceeds that of long distance traffic and thus also accounts for the greatest share of turnover.[501]

[501] In 1929 transport in 2nd and 3rd class accounted for 98% of turnover in passenger traffic with an average distance per person of less than 30 km, see Statistisches Jahrbuch für das Deutsche Reich 1933, p. 163.

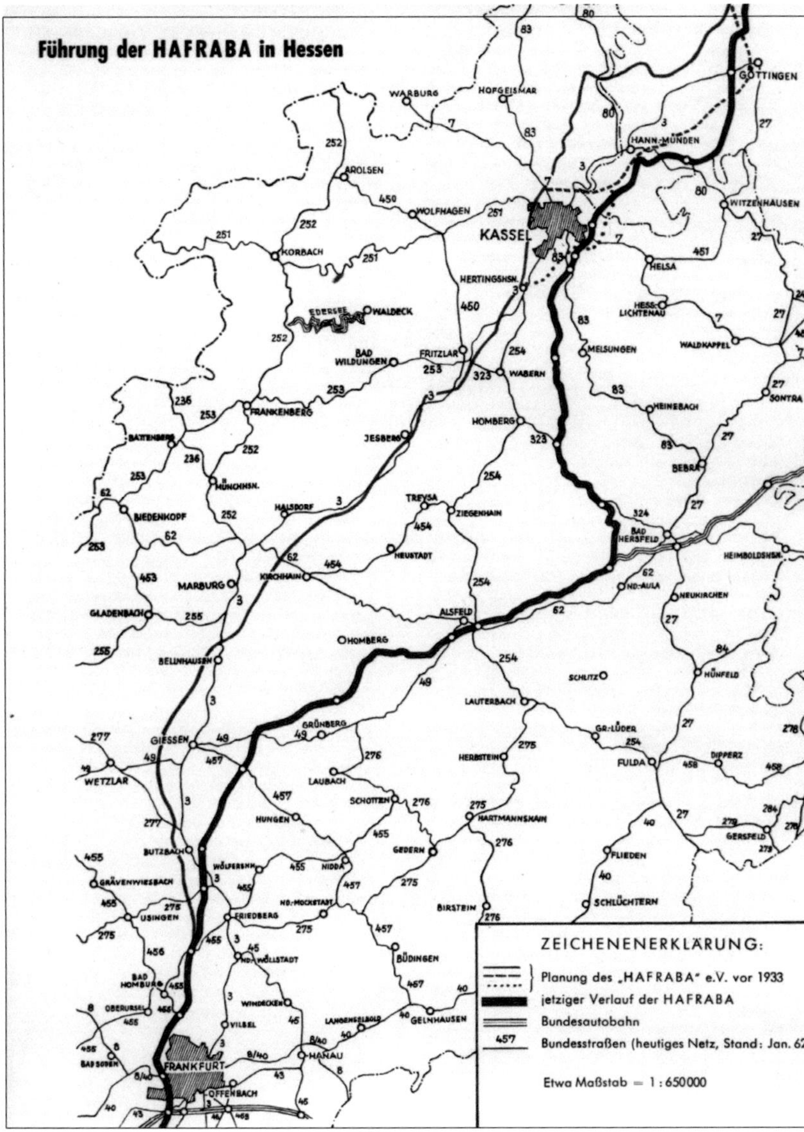

Image 43: The original Hafraba route in Hesse[502]

502 Taken from: Bundesminister für Verkehr (Publisher): HAFRABA – Bundesautobahnen Hansestädte–Frankfurt–Basel, Bau Verlag, Wiesbaden and Berlin 1962, p. 99.

The route network of the A5/A7 and A3 *autobahns* not only shows how little the *autobahn* network was integrated into local transport structures, but its routing also includes some extreme gradients and gives the road the appearance of having been built as an obstacle to truck traffic. The trucks of those days had only a low engine output of some 50 HP. An undated report about the section of the *autobahn* between Dresden and Nuremberg by a Dresden-based road trucking company, concedes that using the *autobahn* cut the time required for the journey by 40% and fuel consumption by 20% compared to the route over the country road. However, the long inclinations of up to five kilometers were a disadvantage and made high demands on drivers, the vehicle, the engine and transmission. The report states: "The uphill section regularly has to be traveled in first gear, whereby on occasion an articulate truck needs up to half an hour to overcome a single inclination."[503]

One can therefore safely assume that fully loaded trucks made only very slow progress on roads with gradients of 7% to 8% and this kind of routing put trucks at a disadvantage to compete with the *Reichsbahn* in long distance transport. Finally, the logical linking of the two forms of transport, truck and rail, that could have been achieved by routing the A4 close to what was then the important rail junction Bebra, was not carried out. There is no evidence in the records of the *Bundesarchiv* that the Reichsbahn exercised any direct influence on routing. The main construction directorates, made up of *Reichsbahn* officials, only carried out the detailed planning of the routes previously laid down by the Gezuvor or Todt.

---

503 BA R 4601, file 3093.

20 The Route Frankfurt a.M.–Kassel–Göttingen

20.1 The Discussion on Transport Policy in Kassel 1920 to 1935

When looking at the geographical location of the city of Kassel in the Federal Republic of Germany, its central location stands out immediately. The question asked now is how Kassel benefited from its location and how transportation concepts in Northern Hesse were implemented. This question can be categorized into the theoretical context of economical geography, which requires the correlation of economic growth and access of a region to the selling market – a question that Adam Smith already alluded to using the example of water transport.[504]

Railroad construction in Germany already took place in the mid 19th century and had to conform to the borderlines of the German small states prior to 1871. For this reason, Kassel was not included in the important north-south and east-west connections. The transportation researcher Otto Blum already pointed out in 1924 that the center of the railroad system in the German Empire of the 1920's lay in the region of Eisenach–Kassel.[505] Kassel, however, was not able to benefit greatly from this since it was not connected to the main axes of the railroad system. There were often complaints about Kassel being difficult to reach by rail, as the predominantly used north-south connection ran through Bebra instead of Kassel. The important connection to the industrial region of Cologne–Dortmund was also poor. The straight-line distance Kassel–Cologne is only 180 kilometers, whereas the railroad distance via Arnsberg is 276 km, via Altenbeken 291, and via Giessen even 301 km. The acquisition of raw materials from the Rhenisch-Westphalian industrial region and the industrial sales in Kassel were impeded due to the poor rail connection. It was Kassel's role as a royal residence city that was a factor hindering the founding and operation of industrial businesses in the 19th century[506], and the unfavorable cargo location hindered its further industrial development.

---

504 The Oxford Handbook of Economic Geography, Gordon L. Clark (ed.), Oxford, Oxford Univ. Press, 2003, in addition: Adam Smith: The Wealth of Nations, Reprint, Bungay, Suffolk 1970, p. 122.

505 Otto Blum: Eisenbahngeographie, in: Erich Giese et al.: Linienführung, Berlin 1924, pp. 1–96, here p. 88.

506 Ingeborg Kirchner: Die Industrielle Entwicklung der Stadt Kassel 1866–1914, Diss. Cologne 1921.

The railroad connection from Kassel to the North was also poor. Only one direct train to Hanover was offered during the day in the 1920s. In order to reach Göttingen, passengers often had to transfer in Eichenberg.[507] The same hindrance existed on the southern Kassel–Fulda route, where passengers had to change trains in Bebra.

In the 1920s, the discussion on transport policy in Kassel was led by the freelance city planner Fritz Stück.[508] In a series of 50 one-page articles in the *Kasseler Post* entitled „The Greater Kassel of the Future“, he introduced recommendations for the development concept of the metropolitan area of Kassel. The last article appeared on 29 January 1928. Stück's proposals were based on the geotechnical approach he had developed. This approach categorized Kassel into increasingly larger areas by increments and led to Stück's view of the Kassel region as the point of intersection between the two global axes London–Tokyo and Hamburg–Cape Town. He pointed to Kassel's inadequate integration in the railroad system. In addition, Stück planned a panorama road 23 meters in width around the Kassel–Fulda Valley with a length of 45 kilometers. This road „creates an organic connection between the green area of the valley and the wreath of the woods. It also emphasizes and secures the crestlines, fields of vision and axes. This monumental frame would create a picture of such overwhelming beauty that it would be unequaled in its uniqueness“, Stück stated.[509] In his writing „On the Future of Kassel“ published by his geotechnical publishing house in 1930, Stück documented the variety of his activities regarding city planning and regional reform in northern Hesse. He used lectures in the Hesse State Museum, radio broadcasts and newspaper articles for his documentation. Among other issues, he addressed the issue of promoting tourism to the nearby Sababurg Castle and the surrounding ancient oak forest, for example. He also referred to the idle estate of Gut Beberek. Stück often bemoaned the lack of transportation in the region of Kassel („Gobi Desert “) and predicted strong growth of the cities of Eschwege and Bebra at Kassel's expense.

---

507 Winter schedule of the Reichsbahn 1930, Kasseler Post 4 October 1930.

508 Folckert Lüken-Isberner: Fritz Stück und die Stadtplanungsdiskussion im Kassel der Weimarer Republik, in: Hessisches Jahrbuch für Landesgeschichte, vol. 40, 1990, pp. 219–240, here pp. 237–239.

509 Winter schedule of the Reichsbahn 1930, Kasseler Post 4 October 1930.

Stück was unable to integrate himself in the social hierarchy of Kassel and earned a reputation as an unpopular admonisher and many of his ideas were ignored, His unfortunate role continued when he was appointed landscape attorney by Fritz Todt in 1935 at the beginning of the NS expressway project in Northern Hesse. He was constantly in conflict with the construction management of Kassel and could not adequately carry out his advisory function. Construction management officials mockingly called him „grass attorney“, and denied him access to the construction sites until the *Reichsautobahngesellschaft* in Berlin intervened on his behalf. Stück finally lost his position as landscape attorney in 1937.[510]

The low level of traffic in Northern Hesse that Stück addressed can be seen in the traffic statistics for 1925. In Kassel, 655 passenger cars, 445 commercial vehicles and 629 motorcycles were registered.[511] Kassel's state of underdevelopment in regard to motorization is especially apparent in comparison to the city of Frankfurt a.M. as seen in Table 7 for the two periods 1927 to 1932 and 1932 to 1938.[512] The number of passenger vehicles and buses per 10 000 inhabitants in Kassel was consistently only two-thirds of that of Frankfurt. The growth of passenger cars (as well as motorcycles) in Kassel remained behind Frankfurt in the first period but then surpassed Frankfurt in the second period without reaching the motorization density of Frankfurt, however.

---

510 Thomas Zeller: Straße, Bahn, Panorama – Verkehrswege und Landschaftsveränderung in Deutschland 1930 bis 1990, Diss. Munich, Campus Verlag, Frankfurt 2002, p. 100s.

511 Armin Schmidt: Die Verkehrsentwicklung Kassel während der letzten 100 Jahre, Diss. Göttingen 1927, p. 35.

512 Statistische Jahrbücher für das Deutsche Reich, 1928–1938.

Year	Passenger cars and buses	Passenger cars and buses per 10,000 residents	Motorcycle	Vehicles in total	1 vehicle is allotted to ... residents
Frankfurt					
1927	4278	90.2	3655	9484	50
1932	7912	148.0	6411	16705	32
1938	18542	334.4	11226	34658	16
Increase 1932–1938 in %:	134.4		75.1	107.5	
Increase 1927–1932 in %:	84.9		75.4	76.1	
Kassel					
1927	1131	65.6	767	2329	74
1932	1943	111.5	1155	3630	48
1938	5644	275.4	2851	10248	20
Increase 1932–1938 in %:	190.5		146.8	182.3	
Increase 1927–1932 in %:	71.8		50.6	55.9	

Table 7: Motorization in Frankfurt and Kassel between 1927 and 1938

As „royal residence" cities one would expect to see similar economic and social structures in these cities. But a comparison clearly shows the underdevelopment of motorization levels in Kassel. Table 8 gives data for the passenger car density of selected royal residence cities in 1929.[513]

City	Passenger cars per 10000 residents
Braunschweig	147.9
Karlsruhe	118.2
Mannheim	117.8
Wiesbaden	107.1
Kassel	104.7

Table 8: Passenger car density of royal residence cities in 1929

513 Statistisches Jahrbuch für das Deutsche Reich 1930, p. 160. Data from Darmstadt is not available, since Darmstadt cannot be considered to be a large city (less than 100 000 residents).

In the low-traffic region of Northern Hesse, the Herkules Hillclimb Race, announced by the Automobile Association of Kurhesse in the 1920s, received a lot of attention and was one of the most important motor sport events in Germany. The objective of the hillclimb was to give the young German automotive industry a platform to test and compare their products, and to „solicit new enthusiasts for the city of Kassel throughout the country."[514]

Despite its character as a royal residence city, Kassel was able to boast several important industrial firms in the 1920s, in addition to harboring the headquarters of the area's iron ore and brown coal mines. Salzmann & Behrens (for the production of linen and cloth, important for the military), the paper mill Papierfabrik AG, the iron foundry Uhlendorf in Bettenhausen, wagon construction (the firms Wegmann and Credé), and the precision engineering and optical industry (e.g. Breithaupt) were all Kassel-based businesses.[515]

Henschel was at the heart of the transport businesses in Kassel with a product spectrum covering locomotives to commercial vehicles. By 1926, Henschel had been a family-managed business for six generations and was considered to be one of the largest manufacturer of locomotives on the European continent and the most important industrial production site in Kassel.[516] Gottwald describes the new business politics of the company's head, Oscar Robert Henschel, as follows: „After World War I and the end of inflation, the sales of locomotives in Germany drastically decreased. For this reason, the young entrepreneur O. R. Henschel was on the lookout for new business fields for his factory and its workers: commercial vehicles, buses, diesel motors and road rollers were manufactured in Kassel as well as steam snow blowers and their first own motor locomotives."[517] An aircraft factory in Berlin complemented Henschel's product range from 1933 on.

---

514 Martin Walter and Heinz Jordan: Das Herkules-Bergrennen in Kassel, 2005, p. 9.

515 Friedrich Frhr. Waitz von Eschen: Der nordhessische Braunkohlenbergbau 1578 bis 2003, in: Zeitschrift des Vereins für hessische Geschichte und Landeskunde, 110 (2005), pp. 113–126.

516 Ingeborg Kirchner.

517 Alfred Gottwaldt: Neuartige Dampflokomotiven von Henschel in Kassel. Das Lebenswerk des Ingenieurs Richard Roosen. Ein Beitrag zur Industriegeschichte Nordhessens 1925–1957, Zeitschrift des Vereins für hessische Geschichte und Landeskunde, 106 (2001), p. 299. See also Kurt Ewald: 125 Jahre Henschel, Kassel 1935, pp. 134s. The celebratory publication by Ewald is specifically technically-oriented. Business data on Henschel, such as turnover or employees is not found.

Image 44: A Henschel Truck in 1925[518]

The firm also produced rail buses that glided on a narrow-gauge railroad such as the one connecting Grifte–Gudensberg in Edertal 1931.[519] Henschel had to greatly draw down its production during the world economic crisis from 1929 to 1933. Henschel was able to bridge the gap in the first year of crisis with an order of eighteen express locomotives from the *Reichsbahn* and 70 construction orders from an emergency program. But the turnover sank dramatically in 1932 to only 47% of the previous year. Up to 1935, Henschel produced altogether 4600 commercial vehicles. The business reports from 1937 show that the Henschel firm achieved half of its turnover in the mid 1930s with the production of the light and heavy trucks that played an important role for rearmament efforts. These were equipped with motors that now had to meet the demand of carrying a permanent load on the new expressways. In addition, Henschel built three-axeled buses designed for long-distance transport with 60 seats – these exhibited a 250 PS motor with 12 cylinders.[520]

---

518 Baumann, p. 66.

519 Ewald, op. cit., p. 240, p. 148.

520 Kasseler Post, 21 July 1937, 15 August 1933, Verkehrstechnik, issue 1, 1930, p.14. Otto Krahl: Strassen entstehen ...Henschel hilft bauen, in Henschel-Stern, issue 2, 1936, p. 27. Kassel City Museum. Krahl was the director of the Henschel Werke in Kassel. For the sector of locomotive production see Bönig, pp. 551–561.

From 23 August to 2 September 1933, the so-called "star ride" ralley to the Tannenberg Memorial in East Prussia took place beginning from locations throughout the Reich on the occasion of the 19th anniversary of the Battle of Tannenberg. Over one thousand automobiles and 483 motorcycles from nation-wide NS-formations and automobile associations participated. The ADAC group in Kassel comprising 500 members also delegated participants. The significance of the star ride was inflated into a huge media event by the Propaganda Minister Goebbels and marketed as the „Eastern Journey of Loyalty". This fulfilled two objectives: promoting vehicle motorization, as can be seen in the discounted gas vouchers offered for this rally, and secondly, political propaganda. On 27 August 1933, Hitler appeared at the Tannenberg Memorial with the aged Reich President Hindenburg in order to partake in his prestige as the victorious military leader of the 1914 Battle of Tannenberg. A tour group from the Transport Association of Hesse and Waldeck, under direction of the transport director Dr. Rehorn of Kassel, participated in this journey as emissaries. One of the Henschel buses described above was painted in the colors of the Reichspost, decorated with swastikas and loaned to the group by the Reichspost. Image 45 shows the bus furnished with the Eastern journey slogan and Henschel advertisement.[521]

Image 45: The Henschel bus for the „Eastern journey of loyalty" (Ostland Treuefahrt)

The group traveled a total of 2300 km. They drove to Weimar on the first day, to Leipzig on the second, and after a 12-hour ride covering 430 km, reached Deutsch Krone in Posen-West

[521] Ewald, op. cit., p. 155. compare also Die Autobahn, issue 1, 1934, p. 17.

Prussia. The journey's length of twelve hours gives an indication of the average speed reached on this trip. On the fourth day the group reached Allenstein in East Prussia via Jastrow, through the Polish Corridor – making a detour to the Reich President's country estate "Gut Neudeck" where Hitler had arrived. Three Schwalmer girls – in traditional dress – who belonged to the traveling group presented flowers to Hitler.[522] Retrospectively, Krahl, director of the Henschel works, came to the conclusion that the high-performance Henschel bus attracted much attention on its „Eastern journey of loyalty".[523]

Kassel was only briefly connected to the air traffic system. In 1926, regular service began at the Waldau Airport which had been built two years earlier. In July 1927, Lufthansa statistics for Kassel Airport showed 487 arriving and 478 departing passengers. In 1928 this number increased to 1470 arriving and 1487 departing passengers with a total of 1574 departing flights.[524] During the world economic crisis of 1929–1933, the city ran out of funds to subsidize air transport. In addition to the sum of 231 000 RM that was needed to subsidize Lufthansa, funds to cover the cost of airport maintenance were also needed. The national and international flight connections (Prague, Rotterdam) were canceled. The airport was nevertheless kept open for large events. A zeppelin, for example, landed in Kassel for the first time on 3 September 1930 and attracted the unfathomable number of 100 000 spectators.[525] The following photo, in which the silhouette of the city can still be recognized, shows the landing in Kassel-Waldau.

---

522 Kasseler Post, 15–29 August 1933, 21 July 1937, p. 17, Verkehrstechnik 1930, issue 1, p. 14.

523 *Krahl*, op. cit.

524 Statistisches Jahrbuch für das deutsche Reich 1929, p. 163.

525 Hessisch Niedersächsische Allgemeine, 100 Jahre Kassel, online Version, 1924–1930, Kasseler Post, 4 September 1930.

Image 46: Zeppelin landing in Kassel-Waldau September 3, 1930 (source: Kassel City Museum)

The following photo shows the zeppelin over the Henschel Villa in Kassel:

Image 47: The zeppelin over the Henschel Villa in Kassel September 3, 1930 (source: Kassel City Museum)

The central location of Kassel in itself was indeed not favorable for air traffic within Germany since the medium-length distances were too short. But before the expressways were built and before fast rail connections could be constructed, air travel was an attractive offering.

Upon the occasion of the „*Wesertag*“ (Weser Day) on 26 October 1934, the *Kasseler Post* published an essay on, „Kurhessen in the Central European traffic region“ written by Dr. Kurt Finkenwirth, legal advisor for the Chamber of Commerce and Industry in Hanover. The author went into great detail on the geographic conditions of Northern Hesse, described the existing transport policy problems there and demanded, among other things, a complete canalization of the Weser and Fulda up to Kassel. The author emphasized that the unfavorable topography in Northern Hesse would be easier to overcome now – with the help of modern expressway technology – than was possible 100 years ago when the railroad system was put into place. He informed the readership that the expressway route Cologne–Kassel–Halle–Leipzig that had been examined by the Hanover–Kassel division of the Gezuvor project company, was revealed to be economically promising based on discussions with the involved Chambers of Commerce. The author concludes his essay with Fritz Stück's observation that Kassel–Kurhessen has a good location in close proximity to the intersecting point of three highly important central European traffic lines whose development into major international travel routes „will lead to the future“. In the 1930's, discussions about the traffic routes and intersecting points were widely spread. The same author, Finkenwirth, wrote a similar study focusing on the city of Hanover.[526]

20.2 Plans for the HAFRABA in Hesse and Kassel

The HAFRABA bulletin published a portrait series of the cities that lay on the Hafraba line. These were largely self-portrayals by the various tourist bureaus.[527] The cities of Hamburg, Kassel, Darmstadt, Mannheim, Heidelberg, Mainz and Bremen were introduced. Even Bad Nauheim, the world-famous springs for treating heart diseases stated that German tourism desperately required „car-only roads“. The director of the Kassel Department of Tourism, Dr. Rehorn – member of the transport committee of the Hafraba since 1929 – published a portrait

526 *Kurt Finkenwirth*: Verkehrszentrum Hannover, in: Hannoversche Woche, 17 July 1930, pp. 5–12, Historical Museum Hanover.

527 Hafraba Mitteilungsblatt, issues 3–8, 1930.

of the city of Kassel in which he highlighted its industries and its location at the intersection of transport routes running from Westfalia to Thuringia and from the Rhine-Main region to Northern Germany.[528] He pointed out that the economic structure of Kassel was marked by a mixture of industry, commerce and tourism that had given the city a healthy economic basis in the confusion of the post-war era. He also addressed the problems of the poor connection to the rail system and stated that the construction of an efficient express train route between Cologne and Leipzig via Kassel had been recommended for 1924. Due to the heavy financial burden for *Reichsbahn* repairs however, this project would not be executed in the near future, nor would a route be realized for north-south traffic from Bremen across the Weser River to Kassel with a workable water way to the Main and Donau Rivers (North Sea–Black Sea–traffic).

Rehhorn pointed out that a planning office had been created to coordinate urban and industrial development for the entire Fulda Valley. The bus lines in the Kassel–Fulda Valley were coordinated in a similar manner from one office. „Thanks to this planned economy, we already have a well-organized and, if necessary, expandable bus system that extends like a spider web and clearly outlines Kassel's needs." To promote tourism, together with the tourist association of Hannoversch Münden, Rehorn implemented the operation of a bus line belonging to the *Reichspost* that traveled from Bremen to Hannoversch Münden and Kassel via Minden. It began operating in 1930 and the travel duration was approx. ten hours at a price cheaper than a second class train ticket but more expensive than a third class train ticket.[529]

In his portrait of Kassel, Rehorn also introduced the planned, odd expressway bridge over the Wilhelmshöher Allee with a photograph taken from the display board of the Hafraba traveling exhibition (cf. Image 13). The following depiction shows this photo montage of the expressway bridge over the Wilhelmshöher Allee.

---

528 Hafraba Mitteilungsblatt, issue 7, 1930, p. 3.

529 Kasseler Neueste Nachrichten, 14 February 1930.

Image 48: Expressway bridge over Wilhelmshöher Allee

Rehorn also referred to his own leaflet entitled „West-East: The great, lively travel route through the heart of Germany“ from the Rhine to the Oder.[530]

The modern traffic concept of the expressway was to be implemented in Kassel as well but the senior planning officer, Kurt Becker, did not handle the issue very cleverly. He presented his concept in newspaper articles in the *Kasseler Tagesblatt* on 31 August 1927 and in the *Kasseler Post* on 1 September 1927. Kurt Becker reiterated his concept in his lecture on 27 January 1928 on the occasion of the opening of the Hafraba exhibition in the Hessian State Museum in Kassel. The Mayor, Dr. Stadler, had invited 80 representatives from public life in Kassel including the planners Stück and Klipp. Hafraba attendees included the speaker of the board Professor Robert Otzen, member of the board Kurt Becker and the managing director Willy Hof.[531]

530 Hafraba Mitteilungsblatt, issue 7, 1930, p. 5s.

531 Invitation list to the opening of the Hafraba exhibition, and Kasseler Post, 29 January 1928, SAK file 10/3.

Kurt Becker wanted the expressway to pass through the Kassel district of Wilhelmshöhe and triggered a considerable storm of protest. His concept envisaged the expressway to pass through the Prinzenquelle and then cross the Wilhelmshöher Allee at the level of the Baunsbergstrasse with a high bridge. This is depicted in the following photo montage.[532]

Image 49: Wilhelmshöher Allee with expressway bridge

This approach also accommodated touristic concepts. The automobile drivers were to be led as close to the Wilhelmshöhe Castle as possible to highlight Kassel's tourist appeal. Kurt Becker referred to experiences with the northern Italian expressways in claiming that expressway traffic was completely noiseless and would not impair the quality of life in Kassel-Wilhelmshöhe. The background to this odd routing of the expressway was also based on the fact that the expressway would advance from Göttingen and already have crossed the Weser between Vaake and Bursfelde before climbing to a height of 380 meters into the Reinhardswald Forest. Its approach from the Reinhardwald Forest led to the preference of bypassing Kassel to the west instead of to the east through Bettenhausen, cf. Image 43. The Hafraba route was to continue to Marburg via Fritzlar. This concept was opposed in many

---

532 Kasseler Tageblatt 31 August 1927.

press articles and letters from the citizens of Kassel. The Civic Association of Kirchditmold, for example, protested against this plan in a letter dated 1 October 1927.[533] Rightly so, noise and exhaust fumes were expected in Wilhelmshöhe as well as the separation of the Wilhelmshöhe Park from the city by the expressway.[534] Since the Hafraba project would be financed with private capital from abroad and refinanced with toll fees, the common assumption was that the international capital providers wanted to increase their return on investment at the expense of the most valuable pieces of property in Wilhelmshöhe.[535]

Fritz Stück agreed with Kurt Becker's ideas in an article in the *Kasseler Post* on 6 November 1927. He also claimed that the route through Wilhelmshöhe was justified and would boost tourism. While Stück presented his approach without thoroughly examining the alternatives, the Kasseler planner Klipp went about things differently. As an alternative, he suggested to design the eastern bypass around Kassel via Bettenhausen and proposed essential elements of the route Göttingen–Kassel–Homberg that were later realized by the NS regime. One advantage of this route was that it had to reach a height of only 320 meters instead of 380 meters. An additional, essential advantage was that the distance of the route between Göttingen–Kassel was shortened by 18 kilometers.[536] The Werra River crossing at Hannoversch Münden had already been planned and the city of Hannoversch Münden was actively involved in the project. The stretch was optimized one more time in agreement with the cities of Göttingen, Hannoversch Münden and Kassel in the years 1930/1931. The agreed upon route was very similar to the one that was realized in the years 1935 to 1937. A maximum incline of 5% was specified.[537]

The Kassel–Fritzlar–Fulda stretch that was stipulated in the STUFA plan no longer appears, however, in the German Reich's system of major roads as it was supplanted by today's B27

---

533 SAK, file 10/3.

534 Schultze: Hafraba und Wilhelmshöhe, in: Kasseler Post 27 January 1928, H. Küster: Hafraba und Wilhelmshöhe, in: Kasseler Post 10 February 1928, SAK, file 10/3.

535 Küster, op. cit.

536 Kasseler Post 3 February 1928, SAK, file 10/3.

537 Landesoberbaurat *Kurt Becker*: Göttingen–Hann. Münden–Kassel, in: Hafraba Mitteilungsblatt Nr. 2, 1931, pp. 1–4.

between Göttingen–Bebra–Fulda, B3 between Kassel–Fritzlar–Marburg and the B83 connecting Kassel–Bebra.

20.3 Expressway Construction in Northern Hesse

The German NS mega *autobahn* project also radiated out towards Northern Hesse. The project draft drawn up by Klipp during the NS planning of the expressway route Göttingen–Kassel–Bad Hersfeld was actually implemented. A special justification for the expressway network around Kassel that was planned during the NS era was its very poor integration within the *Reichsbahn* lines.[538] This explains why the stretch of the expressway Kassel–Göttingen was built first in Northern Hesse, where traffic was actually low. On 1 May 1934, the Main Construction Office for Motor Vehicle Roads was halfway set up in Kassel in the Kronprinzenstraße. Its director was the the *Reichsbahn* official, Liebetrau. It comprised nine departments for routing, real estate purchase, finances and human resources, and two route departments. In addition, there was a department for bridge construction, road pavement and soil science. Construction work was divided amongst the five construction departments in Göttingen, Hannoversch Münden, Bettenhausen, Wilhelmshöhe and Bad Hersfeld. In 1934, the Main Construction Office had 47 public servants, 103 employees and 84 workers, while the five departments had 162 public servants, employees and workers. Companies commissioned to conduct work engaged 3200 workers in the sector of the Main Construction Office Kassel and sometimes worked in a two, or even three shift, system as was done in the construction of the Asbach Valley Bridge near Bad Hersfeld and the Werra Valley Bridge near Hannoversch Münden.[539] The files from the Main Construction Office Kassel are no longer available as they were destroyed in the war.

In sharp contrast to the optimistic expressway propaganda, the working conditions at the expressway construction sites were in many ways inhumane.[540] The workers were conscripted by the employment bureaus from Hesse, and later on from the entire Reich, and had no other alternative due to the bad economic situation. The rearmament program greatly improved the

---

538 Carl Ronsdorf: Hessen und Kassel im Verkehr der Zukunft, in: Kasseler Post 21 October 1934, SAK, Akte 10/3.

539 Kasseler Post 1 May 1936, Susanne Hohlmann, op. cit., p. 42.

540 With regards to the following, see Schütz and Gruber, l.c.

situation on the job market beginning in 1936. As a result of the emerging labor shortage beginning in 1936, the camps for expressway workers were improved in order to provide more attractive accommodation facilities. There is conspicuous news coverage during this time on the supposedly generously equipped camps and picture post cards depicting the work camps were made available for propaganda purposes. Upon the opening of the route Kassel–Göttingen on 20 June 1937, for example, a full-page report appeared in the *Kasseler Post* portraying the tranquil life of the expressway workers on a work-free Sunday at the Knüll Camp near Bergshausen. The following picture shows a postcard of the Holtensen Camp near Göttingen covered with swastikas.

Image 50: Propaganda postcard from the camp in Holtensen near Göttingen[541]

In order to alleviate the hard luck of the camp residents, the organization „Strength through Joy" (Kraft durch Freude) provided entertainment programs to the expressway work camps. Each camp received three theater performances a month.[542] Hohlmann identified six camps in

541 Source: Göttinger Tageblatt 31 May 2007, p. 12.

542 Kraft durch Freude, Gau Kurhessen, Program Februar 1937, p. 39.

the Bad Herzfeld region alone that were transformed into camps for the forced laborers of armament production after 1939.[543]

A different approach to the expressway project than the picture of the working conditions depicted here, is obtained with its oral history, which Britta Eichner-Ramm tracked and recently brought forward.[544] Using the oral history approach, she collected biographical data in the villages along the expressway stretch Kassel–Göttingen. She shows how the expressway construction project penetrated the social system of the villages. It gave the villages new possibilities to earn money by the transport of construction material with horse-drawn carts to the construction site or by providing rooms to the workers as sub-letters. The inns achieved a higher turnover and the county registered increased beer tax revenue. Women found work at the expressway camps preparing food for the workers. The canteens for the construction workers were open to the public, as were the films shown there in the evenings. The village families viewed the construction of the expressway as an event. On the weekends, it was common to visit a construction site and have a morning beer in one of the canteens. The evening film showings brought the new medium „film“ close to the villages for the first time. As a new source of earnings and providing a setting for new experiences, the expressway project was positively assessed by the villages involved.

Still in November 1934, construction of the stretch Kassel–Göttingen began with a ceremony near Sandershausen on the outskirts of Kassel, while in Göttingen the *autobahn* groundbreaking took place at the Kasseler Landstrasse already on 11 September 1934. Construction of the middle section near Hedemünde didn't begin until the fall of 1935 since a final route determination had been delayed because of the difficult terrain. This stretch is marked by its three large bridge structures, the Niestetal Bridge, the Rinderstall Bridge and the Werra Valley Bridge, and involved up to 96 construction enterprises and a peak number of 30 000 workers.

---

543 Susanne Hohlmann: Pfaffenwald – Sterbe- und Geburtenlager 1942–1945, Nationalsozialismus in Nordhessen Vol. 2, Kassel 1983, p. 30. For more information on the camps in the area of the Main Construction Office Kassel see also the interview with the director, Liebetrau, in the Kasseler Post 19 June 1937. Liebetrau names 11 camp locations.

544 Britta Eichner-Ramm: 70 Jahre Autobahn Göttingen–Kassel, Zeitzeugen berichten, Göttingen, 2007.

The monumental bridge structures cemented the claims of the Third Reich as an efficient and powerful regime. The Werra Valley Bridge, connected to a steep grade of 7.8% near Hannoversch Münden and the Fulda Bridge near Bergshausen at the intersection in southern Kassel, were designated as the largest bridge structures of the expressway system by Liebtreu, the main construction office director in Kassel, in a newspaper interview on the occasion of the opening of the Kassel–Göttingen route.[545]

In his chronicle for the Henschel-Stern, a newspaper for the workers of the Henschel Werke in Kassel, Ernst Reyer presents several details about the construction process of the expressway route.[546] For example, to preserve the view into stunning valleys, at two points along the route bridge structures were favored over dams to connect the valleys. The view of Heiligenrode for hikers approaching from Kassel was to be maintained with the help of a large steel structure 160 meters in length. He also highlighted the bridge of the Rinderstall drop, where a deep valley had to be spanned by a bridge with a 30 meter clearance. The beautiful natural surroundings were taken into account and a much larger structure was erected than was technically necessary.

The author goes on the mention another noteworthy bridge structure: the expressway bridge near Ochshausen, where the country road Ochshausen–Waldau, the Söhre rail and the Wahlelbach stream are all spanned by a high-lying structure with a span width of almost 40 meters. The narrative in the Henschel-Stern newspaper noted that the numerous cargo trains needed for the manufacturing of the road surfaces were pulled by Henschel-built locomotives, making the Henschel workers especially proud.

In 1937, after fifteen months of construction in a three-shift operation at a cost of 4 million RM, the bridge over the Werra near Hannoversch Münden was completed. It was considered to be one of the most important structures of its time and was one of the largest structures of the expressway built by the Reich (cf. Image 59). Its height of 60 meters was only exceeded by the 68 meter high Mangfall Bridge in Upper Bavaria and the expressway bridge over the

---

545 Kasseler Post 19 June 19, 1937.

546 Ernst Reyer: Reichsautobahn Kassel–Göttingen, in: Henschel-Stern, issue 4, 1936, pp. 72–76. Stadtmuseum Kassel.

Freiberger Mulde River near Nossen in Saxony with its 71 meter height. Both bridges, however, are 100 meters shorter than the Werra Valley Bridge with 416 meters – the longest and largest in Germany in regard to material used. Headway on the lattice-trussed bridge with an eight meter supporting beam height, is made by two auxiliary supports in a free-standing pre-structure with a distance of 96 meters between pillars. The bridge was made of a rigid lattice construction resulting in a slight bulging of the extremely steep drive up to the Kaufunger Forest, causing the bottom of some trucks and buses to hit the ground.[547]

On 5 April 1945, when American troops had already captured Kassel and were marching towards Göttingen, the Werra Valley Bridge was blasted at the command of the Germans. There were numerous tragedies at the site of the destroyed bridge where people fell to their death because the expressway in front of the bridge had not been sufficiently safeguarded. The old town bridge in Hannoversch Münden, built in 1402, was the only option left for north-south traffic to cross the Werra. The traffic between Göttingen and Kassel was plagued by this solution until a new expressway bridge was completed in 1952.

Image 51: North–south traffic over the old town bridge in Hannoversch Münden in 1952[548]

In 1950, reconstruction of the pillars of the Werra Valley Bridge had resumed and the first half of the bridge was completed by August 1952. Reconstruction costs totaled approximately

547 Eichner-Ramm, op. cit., p. 39.

548 Die Welt on 9 April 1952, Stadtarchiv Hannover.

six million DM. The Philipp Holzmann Company issued an invoice for debris removal in August 1950 for the amount of 112 000 DM to the Road Construction Office of Nordheim. The Werra Valley Bridge now had a tub form and is better suited for the steep hillsides than was the first bridge.[549]

The opening of the route Kassel–Göttingen, on 21 June 1937, began with a grandly staged ceremony at the Kassel East entrance ramp near Heiligenrode, where a grandstand had been especially built for guests. In his speech, the director of the Main Construction Office, Liebtrau, announced that the new expressway decreased the driving time from the Friedrichsplatz in Kassel to the Rathaus in Göttingen by 40 minutes. Gauleiter Weinrich emphasized that the situation of the expressway workers had improved due to numerous amenities: Distance and severance pay had been approved and the NS social services provided the workers with a warm lunch daily. In addition, workers were now issued work shoes and work clothes, which conversely means previously they had been required to to purchase work clothes and shoes themselves.[550] Among those present were members of the military and administrative elite, as well as Oberpräsident (governor) Prince Philipp of Hessen. A convoy of 500 vehicles then rode along the 38 km-long new road to Göttingen. The compact Henschel steam truck with a fuel-oil burner shown below, participated in the ride which also marked the the endpoint of steam-powered technology for road vehicles in the 20th century.[551]

The route was lined with residents of the surrounding villages as spectators. On the festively decorated entrance ramp at Göttingen on the Kasseler Landstraße, where the ground-breaking had taken place on 11 September 1934, a lectern was set up for speakers. After the drive back from Göttingen, a celebration of comradery in a marquee on the Friedrichsplatz in Kassel signaled the end of the opening ceremony. The party newspaper of the NS *gaus* of Kurhessen featured this event between two „parade affairs“ of the NS movement: The Sixth Hessen Day of the National Socialist German Workers' Party (NSDAP) in Kassel and the Reich Fighters'

---

549 Eichner-Ramm, op. cit., p. 39.

550 Kasseler Post 21 June 1937.

551 Carl-Friedrich Baumann: 175 Jahre Henschel, Mores 1985, p. 66.

Day of the right wing Kyffhäuser League in Kassel, thereby connecting the expressway project with the aggressive intentions of the NS regime.[552]

Image 52: Opening ride with Henschel steam truck[553]

Shortly after the route had opened, an expressway service station began operations on 19 August 1937 at the Kassel East entrance. And a change of names at the expressway entrance at Kassel East could be noticed: To the old appellation "Kassel Ost" the designation „Autobahn-Bahnhof Bettenhausen Ost" was added.[554]

The construction work on the route Kassel–Homberg began in the summer of 1935. In the celebratory publication on the occasion of the 600th anniversary of the county of Guxhagen in 1952, the construction of the expressway bridge over the Fulda was described in detail. Image 53 shows a photograph taken from the celebratory publication of the Fulda Bridge at Guxhagen from the period after 1937. The bridge was blasted in the last days of the war and rebuilt as a single-lane bridge by 1952.[555]

---

552 Kurhessische Landeszeitung 19/20 June 1937 and Kasseler Post 21 June 1937.

553 Photograph from a Henschel issue, No. 13, August 1937, p. 92. Henschel Museum Kassel. For the Henschel steam wagon see H. Uhlig: Betriebserfahrungen mit dem ersten Henschel-Dampfautobus, in: Henschel issues, March 1934, pp. 6–8, Henschel Museum Kassel. The steam bus rode more smoothly and quietly on hilly routes than a diesel bus did, which strongly vibrated and required gears to be shifted numerous times.

554 Kasseler Post 21 June 1937

555 Adam, Werner: Entstehung und Bau der Reichsautobahn, in: Festschrift 600 Jahr Feier Guxhagen, Guxhagen 1952, p. 65–71, here p. 69.

Image 53: Expressway bridge at Guxhagen

The opening of the route Kassel–Homberg on 17 December 1937, was part of altogether six opening ceremonies that took place simultaneously throughout the Reich to celebrate the completion of the first 2000 km of expressway. The director of the expressway project, Fritz Todt, held a commemorative speech at the Berlin belt at Erkner/Grünheide that was broadcast on the radio. This made it possible for the Main Construction Office in Kassel to listen to Todt's speech at the ceremonial activities of the service station at Kassel East.[556]

The expressway undoubtedly had a utopian character, since the traffic density in Germany at the time was very low in comparison to other European countries. In addition, Northern Hesse was deemed to have especially low traffic levels. Pictures from that time show expressway stretches without an, or only very little traffic. For photographs and propaganda events, traffic on the expressway had to be especially staged. The following Image 54 shows the expressway without cars at Guxhagen.[557]

---

[556] Kasseler Post 18 December 1937.

[557] Fünf Jahre Arbeit an den Straßen Adolf Hitlers, Berlin 1938, p. 58.

Table 6 lists the opening dates for selected expressway sections[558] and shows how neighboring cities were connected beginning 1933: Frankfurt with Darmstadt, Mannheim with Heidelberg, Leipzig with Halle, Kassel with Göttingen as well as Cologne with Düsseldorf. It is striking in what short amount of time the route Frankfurt–Mannheim was completed, especially in comparison to the much shorter stretch Kassel–Göttingen (38 km) which had a lower priority and was completed almost two years later.

Image 54: Expressway without vehicles at Guxhagen

The following picture shows a propaganda photograph with a view of Kassel from the autobahn junction Südkreuz Kassel.

558 Statistisches Jahrbuch, 1936, p. 198.

Image 55: View from the Südkreuz intersection to Kassel (propaganda photograph).[559]

The expressway project was under the threat of being discontinued due to the changed economic priorities of the war period. The project tied up scarce manpower and required scarce raw materials such as steel and cement without making any contribution to strengthening the German military. For this reason, in complete contrast to the period before 1940 where openings were spectacularly celebrated, the expressway project was to be continued as inconspicuously as possible and without press representatives. The route Kassel–Hanover continued to be built during the war in this manner, for example. When the thirteen km long stretch between Göttingen–Nörten was provisionally completed on August 3, 1942 and opened to traffic, the district administrator of Göttingen withheld public announcement and refrained from informing the press.[560] As a result of the new order of priorities during the war, the construction sites of the Kassel–Eisenach route were abandoned. Only the isolated expressway bridge in the woods at Vollmarshausen near Kassel attests to its one-time existence as a part of a plan. In its stead, the route Bad Hersfeld–Eisenach was provisionally

559 Fünf Jahre Arbeit an den Straßen Adolf Hitlers, Berlin 1938, p. 57.

560 Schwell and Jäger, op. cit., p. 479.

completed on July 1, 1943, in the fifth year of war.[561] The following picture shows the isolated expressway bridge at Vollmarshausen.

Image 56: Isolated expressway bridge in the woods at Vollmarshausen (private photograph taken in 2006)

The construction of the expressway continued until 1943 with forced laborers who lived and worked under the harshest conditions.[562] Rudolf Herz and Reinhard Matz established a connection between the construction of the expressway with forced laborers and the Holocaust when, in 2001, they proffered the idea of paving a one-kilometer long segment of the stretch with stones, and introduced a reduction of the speed limit to 30km/h to remind drivers of the victims.[563] Image 57 shows a photomontage of the planned project.

561 Daten zur Bauleistung according to Dietrich Schwell and Wolfgang Jäger, l.c., p. 485s.

562 Zeller, l.c., p. 64. Susanne Hohlmann.

563 Rudolf Herz and Reinhard Matz: Zwei Entwürfe zum Holocaust-Denkmal, Nuremberg 2001, p. 46.

Image 57: Draft for a Holocaust memorial near Kassel by Rudolf Herz and Reinhard Matz

It was only when the military situation dramatically climaxed after the invasion of the allies in France in June 1944, that the *Reichsautobahngesellschaft* was completely dissolved „due to the *Führer's* edict on the total war mission dated 27 July 1944“.[564] As gas became increasingly scarce during the last years of the war, motorized traffic stood completely still and non-motorized traffic such as pedestrians, bicyclists and animal-drawn carts took over the expressway. Already in 1942, the District Administrator of Göttingen spoke of „an improper use“ of the expressways. Image 58 shows two bicyclists on the expressway near Dahlenrode in 1945.[565]

One aspect of the propaganda effort to be discussed, is the incorporation of art into the expressway project. The large extent to which Todt instrumentalized art for his purposes is surprising. The traveling exhibition "The Roads of Adolf Hitler in Art " (*Die Strassen Adolf Hitlers in der Kunst*) showed the aquarelles and graphics by Ernst Vollbehr in many cities. The Kassel Art Association also showed this exhibition in October 1937, as did the Welf Castle Hannoversch Münden in June 1937.[566] The Vollbehr painting in Image 59 shows the construction of the expressway bridge at Hannoversch Münden. Todt's systematic incorpora-

[564] Bundesarchiv Berlin R2/25018.

[565] Photograph from Eichner-Ramm, op. cit., p. 48.

[566] Kasseler Post 6 October 1937, Göttinger Tageblatt 15 May 2007.

tion of numerous artists into the expressway project clearly shows that he understood the project to be a grandly created propaganda production by the Third Reich.

Image 58: Bicyclists on the expressway in 1945

Image 59: Construction of the expressway bridge at Hannoversch Münden[567]

[567] Ernst Vollbehr: Battle of Labor – 5 Years of On-The-Scene-Painting of the Expressways of Adolf Hitler.

21 Constructing the Autobahn Munich–Salzburg

21.1 Introduction

In the research on the expressway construction by the National Socialists from 1933–1939 little attention has been paid to the historical aspects of the regional planning and construction of the expressways. This chapter examines the expressway Munich–Chiemsee–Salzburg.

From a regional historical perspective, this topic is particularly interesting as some of the major players of the expressway project – including Todt, Seifert and Woerner – operated out of Munich for many years. Moreover, the construction of this route was accompanied by an enormous propaganda effort whose powerful historical evocations reverberated even decades later. Indeed, the legend of the „green expressway“ was created: The reconciliation of technology and nature was proclaimed and the massive intrusion into nature by these major building projects was cloaked by adapting the expressways to the landscape.[568]

In the overall National Socialist expressway project, the concept of the Chiemsee expressway is unique inasmuch as it was planned foremost as a convenient means for reaching the rest and recreation centers of Tegernsee (Lake Tegern) and Chiemsee (Lake Chiem) from the city of Munich (population: 800 000). The factor of rest and recreation did not figure in any other of the German Reich's expressway segments. The Berlin–Stettin expressway which serves to reach the Baltic Sea resorts for the capital's population offers a good analogy, however, the distance of 170 kilometers to Stettin is significantly farther than the distance from Munich to Chiemsee. The only route that is comparable to the Chiemsee expressway is the 36 km-long route from Milan to Lake Como which was constructed in the 1920s by Puricelli – an avowed advocate of the expressways.[569] Here too, the expressway's foremost purpose was for reaching the rest and recreation areas in the vicinity of Milan. It was Puricelli also who played no small part in the development of an expressway concept – together with the Munich-based Sager & Woerner construction company – for reaching Lake Starnberg during this same basic time-

568 The concept of myth of green expressway was introduced by Thomas Zeller in his study „Driving Germany – The Landscape of the German Autobahn 1930–1970“, New York, 2007.

569 Uhlfelder, H.: Die oberitalienischen Autobahnen, in: Hafraba-Mitteilungsblatt, issue 7, 1930, pp. 9–14.

frame (see below). The expressway concept of making rest and recreational areas easily accessible for the large population centers was widely espoused in the 1920s, cf. Section 6.

21.2 The Planning Context

The planning for the construction of the Munich–Salzburg expressway got under way at a time when various road construction projects were being stimulated throughout the German Reich and the rest of Europe.

In the 1920s there had been deliberations to construct the road connecting Munich with Passau as a major road rather than as a secondary route. It would pass through the Vilstal and tap into the Lower Bavarian region where it would allow the forging of a connection to the Donau transshipment port in Passau.[570] A trans-alpine road from Lindau to Bad Reichenhall – which would also connect the popular excursion destinations Reit im Winkl and Ruhpolding – was envisioned by an association that formed to support this cause. This idea was picked up by Hitler in 1933[571] and executed parallel to the Munich–Salzburg expressway project.[572] It was anticipated that the building of an attractive alpine road would discourage automobile tourists from traveling to Austria and Switzerland to enjoy similar roads. In addition, the road connecting Munich, Rosenheim and Bad Reichenhall via Teisendorf was straightened and broadened from 1933–1934.[573] In the following year, the Munich–Garmisch-Partenkirchen route was upgraded in preparation for the 1936 Winter Olympics; a 14 kilometer-long leg of this „Olympic road" was completely reconstructed. Altogether, Upper Bavaria hosted four

---

570 Minutes from a meeting held on 5 June 1925 in Munich with the Chamber of Commerce, Munich and the Bavarian Automobile Club, Stadtarchiv Munich, Verkehr/Transport 84. See also Roland Gabriel: Deutsche Autobahnvorläufer – AVUS, Hafraba, München–Leipzig–Berlin, Leipzig–Halle, Düsseldorf–Köln–Bonn, in: Wolfgang Wirth: Die Autobahn – Von der Idee zur Wirklichkeit, Cologne 2005, pp. 38–40.

571 Münchner Zeitung on 28 March1933, Stadtarchiv Munich, Verkehr/Transport 84. See also Fischer, Josef: Entstehung, Linienführung und bauliche Ausgestaltung der Deutschen Alpenstraße, Die Strasse, issue 7, 1935, pp. 208–212.

572 The alpine-crossroad project was supervised by Todt alongside his main duties as chief civil engineer. Todt reported that approx. 600–700 would find work on the Bad Reichenhall – Inzell segment in the Spring of 1934, Münchener Zeitung on 16 January 1934, SM, München–Salzburg 1933–34. Similar to the expressway project, the alpine-crossroad was also heavily integrated in the Nazi regime's propaganda efforts. In September 1934, for example, the construction sites were visited by Berlin-accredited diplomats before continuing on to the NS party convention, see Die Strasse, issue 3, 1934, p. 92. Tomas Zeller (2009) explored the construction of the transalpine road.

573 Münchener Zeitung on 8 November 1934, SM, München–Salzburg 1935.

major road construction projects that were in progress parallel to one another from 1933 to 1938.[574] This paper, however, will limit its focus to the Munich–Salzburg expressway.

The main road Munich–Rosenheim–Teisendorf was a major route to Salzburg and stretched across a length of altogether 142 km. Image 60 shows the Munich–Salzburg route via Wasserburg.[575] While only 128 km in length, this route passed through the 25 villages that are listed in Table 9 and gives an indication of how time-consuming automobile travel was on this leg. When assuming an average speed of 30 km/h, this trip required 4 hours. Compared to driving on an expressway with an assumed speed of 100km/h, the stretch could be covered in one hour and 12 minutes. In other words – a main road to expressway travel time ratio of 4:1.

The planning process of the Munich–Chiemsee–Salzburg leg was carried out in the following context: Based partially on the circumstance that Munich had a high automobile density, a high volume of traffic to the surrounding areas could be observed on weekends. With one vehicle (including motorcycles) for every 24 inhabitants, Munich's motor vehicle density was Germany's highest in 1932 (compare to Berlin with one vehicle per 38 inhabitants.).[576] Munich's position at the forefront of motorization is demonstrated also by the fact that from the beginning of the 20th century, the General German Automobile Association (ADAC) was based in Munich. In addition to private transport, 13 bus routes offered travel to the excursion destinations around Munich. These bus lines were operated by the *Reichspost* and transported 700 000 people in 1934. According to the *Reichspost*, the most popular weekend destinations were Bad Tölz, Grünwald, Wolfratshausen, Starnberger See, Ammersee and the River Inn.[577] The following Image 61 shows the destinations that could be reached by bus from Munich.[578]

---

574 Sommerer, Wilhelm: Ausbau der Reichsstraße zu den Kampfstätten der IV. Olympischen Winterspiele (Reichsstraße Nr. 2, München–Garmisch) und Verkehrsleistung während der Spiele, in: Die Strasse, issue 5, 1936, pp. 156–157. The construction of this route is based on a project sketch from the Munich-based construction company Sager&Woerner in 1927, compare footnote 22.

575 ADAC (General German Automobile Association) (Publisher) Reichs-Verkehrsführer und Reichs-Tourenbuch, Dortmund 1930.

576 Statistisches Jahrbuch für das Deutsche Reich, 1933, p. 159.

577 Postal service operated bus traffic from Munich, in: Die Strasse, 1935, issue 12, appendix, p. 33.

578 Op. cit.

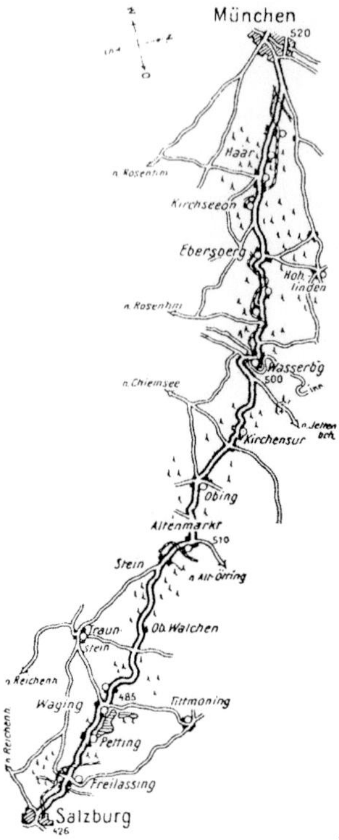

Image 60: Road
München–Salzburg

	Distance to the next village, km	**Distance accumulated, km**
München (Ausfahrt: äußere Wienerstraße)	0,0	0,0
Wegteilung von Zamdorf	2,7	2,7
Strasstrudering	3,7	6,4
Haar	4,6	11,0
Zorneding	8,4	19,4
Kirchseeon	6,3	25,7
Ebersberg	5,2	30,9
Steinhöring	6,6	37,5
Tulling	1,2	38,7
Reitmehring	10,1	48,8
Wasserburg am Inn	3,2	52,0
Strass	3,0	55,0
Kirchensur	7,2	62,2
Frabertsham	3,3	65,5
Obing	4,0	69,5
Rabenden	5,0	74,0
Altenmarkt	5,5	80,0
Stein	2,5	82,5
Ober-Walchen	5,4	87,9
Wegkreuzung	2,0	89,9
Otting	6,3	96,2
Waging	3,0	99,2
Petting	7,0	106,2
Schönram	5,0	111,2
Freilassing	9,4	120,6
Liefering	2,5	123,1
Salzburg	5,0	128,1

Table 9: Villages on the the road Munic–Salzburg

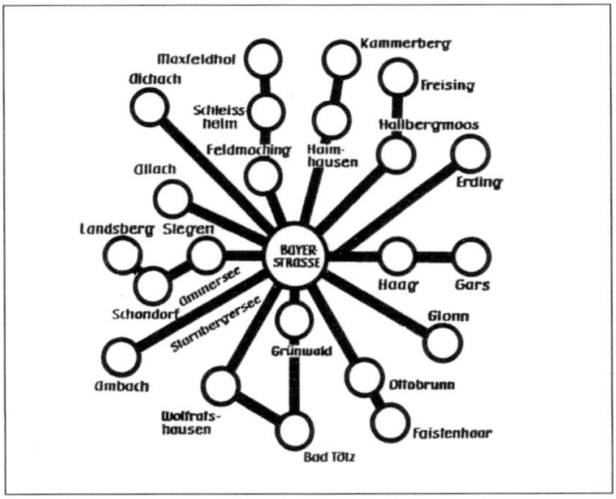

Image 61: Easily accessible destination from Munich with bus lines operated by the postal service

To improve access to these recreation areas, the Munich-based construction company Sager & Woerner – where Todt was employed as supervising engineer beginning in 1925[579] – issued a pilot study on the Munich–Starnberg expessway.[580] According to Nazi press estimations, the study appeared to have been instigated by Puricelli who is alleged to have met with Alfons Woerner, the construction company's owner, in 1926 in Munich.[581] Sager & Woerner justified the plans of constructing an expressway from Munich along the Lake Starnberg's eastern shorefront – with an option of being extended to Garmisch-Partenkirchen – by the anticipated weekend traffic to Munich's popular rural outskirts. The route which was regarded as a „typical sports and excursion road," would serve to access areas that the *Reichsbahn* did not yet service.[582] Image 62 shows one of the planning charts.[583]"

[579] Eduard Schönleben: Fritz Todt, Oldenburg, 1943, p. 30s. See also the company history of Sager&Woerner 1864 to 1964, Volume 2: 1920 to 1945, p.115, Munich, Selbstverlag, 1964 (Bayerisches Wirtschaftsarchiv, Munich), p. 211.

[580] Vilbig: Ausbau der Straßen – Nur-Autostraßen? Drei Musterbeispiele aus Bayern, in: Verkehrstechnik, issue 21, 1927, pp.327–330. Undersecretary Vilbig of Munich participated in Leipzig's 1927 highway convention.

[581] Münchner Augsburger Abendzeitung on 20 February 1934, SM, Autobahn München–Salzburg 1933–1934. In Sager&Woerner's company history from 1864 to 1964,Volume 2: 1920 to 1945, Munich, Selbstverlag, 1964 (Bayerisches Wirtschaftsarchiv, Munich) a meeting between Puricelli and Woerner in Munich took place „around 1924", see p. 212.

[582] Vilbig, op. cit., p. 330, also van Elsen, Paul: Die deutsche Landstrasse, Diss. Cologne, Düren 1929, p. 89.

[583] Vilbig, ibidem.

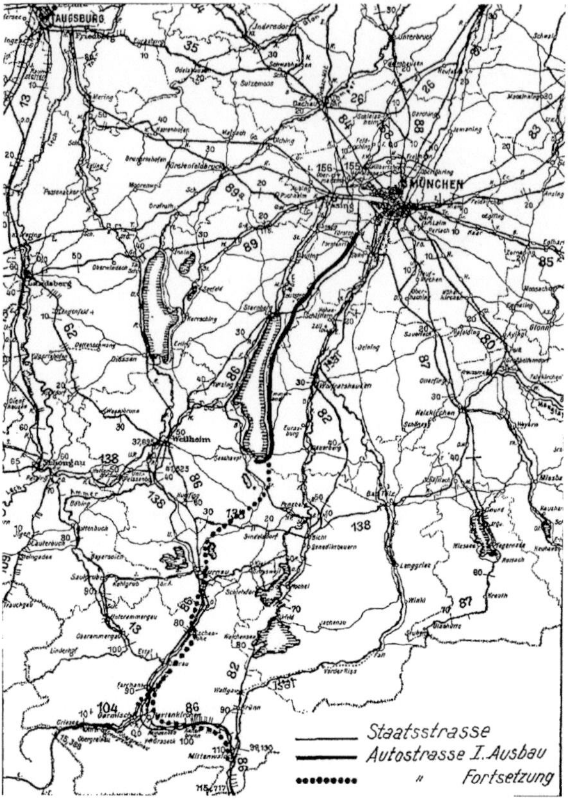

Image 62: Expressway Blueprint of Munich–Starnberg Route from 1927

Compared to the accessible regions that lay to the west of Munich, important tourist destinations east of Munich–Tegernsee or Chiemsee – could not be reached by bus. Woerner described the region as poorly accessible.[584]

Plans for the Chiemsee expressway focused heavily on this anticipated weekend traffic. It was Kaftan's opinion that too many visitors traveled from Munich to Lake Starnberg creating a heavy burden on the local infrastructure. Furthermore, he quotes a source according to which

[584] Anton Woerner: Die Reichsautobahn München–Salzburg, in: Die Autobahn, issue 9, 1933, pp. 10–12, features a short biography and a portrait photo of Woerner.

Todt – in cooperation with tourist organizations and other interested parties – had already compiled a pilot study on a Munich–Chiemsee expressway before 1933.[585]

21.3 Choosing the Route

The extent to which the expressway project had been coordinated with the infrastructural demands of Munich's excursion traffic is easily demonstrated by the selection of the route. Heretofore, there has been little historical research in this direction.[586] This route was cited in numerous articles by the Nazi press as an example of how an expressway could be adapted to the landscape and how tourist attractions could be linked with the building of an expressway. From 1934 to 1938, Todt's journal „*Die Straße*" reported on the Munich–Salzburg expressway 36 times – the journal's precursor „*Die Autobahn*" issued seven reports. In addition, the publishing house "*Volk und Reich*" printed a script entitled „*Die Reichsautobahn München-Salzburg*" (Berlin 1936). The powerful legend of the „green expressway" was created – in part – to conceal its considerable encroachment into the landscape. The claims that the expressway had adapted to the landscape and that engineering had been reconciled with nature were put forth to conceal the harmful consequences of this construction work to the environment. Todt's journal makes no mention of either flora or fauna – even though the construction would put at risk such animals as the mountain cock and the great crested newt. Notwithstanding, this legend did challenge many theorists of landscape architecture to reflect on this issue in later decades.

It can be assumed that there was wide-spread approval for the planned course of the Munich–Chiemsee–Salzberg expressway in politics, administration and population at large. One prominent case of opposition, however, the so-called „Halter affair" is known and was first documented by Schütz and Gruber.[587] Georg Halter, a professor for road construction and railroading at the Technical University in Munich, was a renowned planning expert who had joined the NS party even before 1933. In an apparent state of complete unawareness of the fact that freedom of expression was all but non-existent during the Nazi era, he wrote a series

---

585 K. Kaftan, p. 153.

586 See Claudia Windisch-Hojnacki.

587 E. Schütz and E. Gruber, op. cit., p. 20s.

of critical statements on the expressway project. Todt was extremely agitated and undertook measures to silence Halter; he initiated a meeting between Doll and Halter, for example.[588] In a letter to the professor, Todt invokes the name of Minister Hess – the *Führer's* deputy – with whom he had conferred on this matter, to threaten the professor that „any kind of critical written publications or spoken statements were considered undesirable and irreconcilable with his position as a university professor."[589]

Image 63: Panoramic View of the Plain and Alps near Rosenheim

For one, Halter had objected to the length of the southerly route.[590] In choosing the southerly route over the northerly one, the Munich–Salzburg leg was made longer by 15 km – a circumstance that Halter rejected on the grounds that it was not reconcilable with an efficient use of the funds needed for the long-distance route to Salzburg. Halter was less interested in the

588 Halter released four copies of the memorandum into circulation. Among others, Vice Chancellor von Papen received a copy. See report from Doll, head of construction project in Munich to Todt on 31 January 1934 about a discussion with Halter, Bayerisches Staatsarchiv, Munich, Autobahndirektion Südbayern, 248.

589 Bundesarchiv, file R4601, volume 1307, sheet 100. The votes against the expressway are collected in this volume.

590 Georg Halter: Gedanken zur Linienführung der Autobahn von München nach Salzburg, vom 21.10.1933, ibidem.

project as a means to make tourist destinations more easily accessible from Munich. He also criticized the inconsistency of committing extensive funds for the design of a route that would allow travel speeds of up to 160 km/h around hill tops and in curves in the name of reducing travel time while also choosing the longer southerly route over the northerly route. Halter advocated a more economical use of the financial resources and a route design for lower speeds. As he argued in favor of the northerly route for financial reasons, Halter also proposed funding further expressways with the money saved in a bid to increase the economic counter value of the invested funds.

The connection that Todt established in his writings between fast expressway driving and a simultaneous enjoyment of the landscape was yet another aspect that Halter rejected as contradictory. This was a frontal attack on a central tenet of Todt's expressway ideology – the summons to „*Autowandern*" to enjoy the landscape while speeding along the expressways. In his journal „*Die Straße*," Todt published numerous articles to propagate this idea.[591] These were elaborated with suggestions for picnics on the grass borders along the expressway, cf. Image 22.

Todt tried to attenuate the conflict between fast driving and landscape enjoyment by repeatedly stressing, in press releases at the start of constructions for the first expressway routes of Frankfurt to Darmstadt and Munich to Holzkirchen, that the expressways would not be race tracks but served – above all – the enjoyment of the landscape.[592]

Halter also pointed out that the expressways would be ill-suited for military purposes because the road surfacing could not withstand the weight of the tanks or other heavy artillery equipment. The wide bridge spans of the southern routes could become targets during air attacks and this vulnerability would limit their usefulness for military purposes. Despite Halters concerns, German tanks rolled across the expressways towards Austria in 1938.[593]

---

591 On "Autowandern" see Die Strasse, issues 3, 6, 11, 14, 1936.

592 Münchner Neueste Nachrichten on 24 December 1933, SM, München–Salzburg 1933–1934, Münchner Augsburger Abendzeitung on 21 March 1934, ibidem.

593 See Guderian: Mit der Panzerwaffe auf den Straßen des Sieges, in: Die Straße, issue 23/24, 1940, p. 594: Guderian stresses the high average speeds that are attainable on the expressways while „being gentle on both man and machine": „What joy to march within in old territory of the German Reich!"

21.4 Adaption to the Landscape

In order to adapt the expressway to the landscape, Todt unleashed several special activities for the Munich–Chiemsee–Salzburg leg. Todt's approach to the layout of a road was completely different from the way that a train engineer might be expected to plan a railroad. This, of course, caused some concern about how the business culture of the *Reichsbahn* could be adapted to the requirements of the expressway construction plans once the train engineers were delegated to the planning agency of the *Reichsautobahngesellschaft.* As a corrective measure, Todt mandated landscape advocates to the highest level of the site management where they were to serve as consultants on the adaption of the expressway to the landscape. As contract employees, however, they could exert only little influence. It remains puzzling that Todt did not create permanent positions for them although he deemed their work extremely important.

At the end of 1933, Todt named the Munich-based landscape architect Alwin Seifert, with whom he had a congenial relationship, as the senior landscape advocate.[594] Both men had in common the ideas of a homeland-protection movement that had been initiated in the 1920s as an answer to rampant industrialization and its detrimental effects on the environment. This movement demanded a gentler adaption of factories and industrial-scale technologies to the landscape.[595] Todt was able to integrate his idea of homeland protection within the Nazi ideology by invoking the notion of a „German technology". In his first letter to Seifert in 1933, Todt had already formulated his idea that the expressway in the Hofolding Forest near Munich should be incorporated into the landscape – not like a railroad embankment but rather like a parkway. He elaborated that: „Perhaps it would be possible to heighten the impression of a „park way" through a forest by creating especially level slopes and by a low positioning

594 See convention of the top managers of the Main Construction Offices with Seifert and Todt in Berlin on 18 January 1934 incl. Photo of participants at a reception at the Reich Chancellery, in: Die Autobahn, issue 2, 1934, p. 48.

595 Zeller 2002, p. 74. The course of the railroad lines unleashed much criticism. The Munich–Starnberg route that was inaugurated in 1854 proceeded along the banks of Lake Starnberg and made it inaccessible to the public, see Wolfgang Pusch: Die „Entdeckung" des Würm- oder Starnberger Sees und die Folgen, in: Landpartie. Museen rund um München, published by Landpartie, Munich 2002, pp. 180–195, here p. 189. In a letter to Todt on 12 April 1936, Seifert critisizes the constraints at Lake Constance stemming from the railroad, see BA, Akte NS26/1188.

of the road."[596] In the same letter, Todt asked Seifert to walk through the deforested areas of the Hofolding Forest for inspiration on how best to adapt the course of the road. Todt's request indicates that the felling of the trees along the course of this leg had already been determined and was no longer something that Seifert was able to influence. According to reports from media outlets in Munich, Seifert – as a landscape architect – was able to ensure that groups of trees to both sides were preserved to ensure a park-like character.[597] These articles also reinforced the idea that the concept of landscape aesthetics – and Seifert as the seminal figure to propagate it – formed the cornerstones of the public relations campaign early on. The campaign aimed at garnering acceptance for the project and laying the foundation for the „green expressway" legend.

The scarping of the expressway from the surrounding landscape was, however, not carried out consistently. In a letter to the Berlin directorate of the *Reichsautobahngesellschaft* on 17 October 1935, Todt complained that an inspection of the Holzkirch–Rosenheim leg revealed that instead of having achieved a natural integration of the roadway into the landscape, in numerous cases the roadway had been carved into the landscape via unnecessary ditches; in other cases, it had been cut-out of the landscape via ditches that were unnecessarily deep.[598]

With great enthusiasm for his new task, Seifert inspected all of the important construction sites of the Reich in the following years and communicated his appraisals to Todt.[599] Moreover, between 1934 and 1942 he published a total of 30 articles in Todt's journal „Die Strasse" dealing with landscape design in connection with road construction. In a groundbreaking study, Zeller carefully examined Seifert's influence in the planning of the expressways. Zeller also revealed how Todt adopted the concept of landscape aesthetics from the parkways in the United States.[600]

---

596 Letter from 23 November 1933 , ADM, NL133/056.

597 Münchner Augsburger Abendzeitung on 20 February 1934, SM, ZA München-Salzburg 1933–1934.

598 ADM, NL 133/63

599 The Seifert estate NL 133, Archive, Deutsches Museum, documents travel expense reports.

600 Zeller 2002 and Zeller 2007. Also on the parkways: Bernd Kreuzer: Tempo 130, op. cit., pp. 13–17. Also: Alexander Thewalt.

In the summer of 1934, Seifert undertook several boat excursions on Chiemsee to anticipate the effects of a bracing of the expressway by means of a wall at the lake's edge. He discouraged the building of such a wall in favor of an escarpment toward the lake. „A wall would prove to be too disruptive for the composition of the Chiemsee's shoreline and hinterland. – it would form an unacceptable foreign object“.[601] Image 64 shows the Chiemsee parking lot with a KDF-tour bus.[602] The fact that this wall was built at the Lake Chiem parking lot shows that Seifert was not always able to assert his ideas. The unequal pair Todt and Seifert also appeared in other contexts: Together they prepared the exhibition „*Die Strasse*“ (see below) and also cooperated during the extension of the road to Obersalzberg.[603]

Image 64: Rest Stop with Parking Lot at Lake Chiem with a KDF-Bus

21.5 The Decision Making Process and Beginning of Construction

According to Munich's press corps and Woerner, Todt took the planning of the Chiemsee expressway into his own hands and presented the concept to Hitler at the beginning of July

---

601 ADM, NL 133/56, letter to Todt on 26 June1934.

602 From: Fünf Jahre Arbeit an den Strassen Adolf Hitlers, published by the General Inspector for German roads, Berlin 1938, p. 91.

603 Letter from Todt to Seifert on 1 December 1934, ADM, NL 133/56.

1933; he outlined both alternatives – the north as well as the south route to Chiemsee.[604] The southern route along Tegernsee, Bad Tolz and Bayrischzell was chosen to accommodate the main tourist streams.

Alfons Woerner, owner of the Munich-based construction company Sager & Woerner, was one of the expressway project's important players. Todt had been employed at Sager & Woerner as a chief engineer until 1933. With the inception of Woerner as a member of the Hafraba board of administration on 23 May 1933, Todt had a confidant in this society.[605] At the founding of the expressway project society Gezuvor on 18 August 1933, Woerner was named managing director of the Bavarian section of Gezuvor.[606] Within the same summer, Sager & Woerner were commissioned to carry out a project study for the expressway leg Munich–Salzburg – four of the copies went to Todt and the *Reichsbahngesellschaft*.[607] The close relationship between Sager & Woerner and the powerful Todt may have been the trigger for the company's shabby behavior vis a vis sub-contractors while executing the contract sections for the Munich–Salzburg construction phase. Numerous complaints are filed in the dossiers of the Bavarian State Archives.[608]

In August of 1933, the decisions and planning procedures for the construction of the Munich–Salzburg expressway came fast and furious. Since the Gezuvor comprised eleven regional sub-groups with merely regional responsibilities, the breakup of the Hafraba Association on 18 August led to an expansion of Todt's power.[609] The overall responsibility for the entire network now fell to him. It came as a complete surprise to road planners and transportation

---

[604] Münchner Augsburger Abendzeitung on 20 February 1934, SM, ZA München-Salzburg 1933 to 1934, and Woerner, op. cit., p. 10.

[605] Minutes of a membership meeting on 23 May 1933, Die Autobahn, issue 6, 1933, p. 7.

[606] Die Autobahn, issue 9, 1933, p. 8f, issue 10, 1933, p. 21.

[607] Letter from the directorate of the Reichsautobahngesellschaft Berlin to the Main Construction Mgt. Munich on 13 March 1934 requesting a blueprint and the answer on 20 March 1934, in: Bayerisches Staatsarchiv, Munich, Autobahndirektion Süd, Volume 247. The commissioning of a pilot study is also mentioned in the company history of Sager&Woerner 1864 to 1964, volume 2: 1920 to 1945, p.115, Munich, Selbstverlag, 1964 (Bayerisches Wirtschaftsarchiv, Munich).

[608] Letter on 17 September 1935 from the expressway field office in Rosenheim to the Main Construction Mgt. in Munich, Bayerisches Staatsarchiv, Munich,volume 573.

[609] In his speech in Frankfurt on 18 August 1933 to high-ranking representatives of the Laender on the occasion of the restructuring of Hafraba into Gezuvor, Todt admonished Gezuvor to limit its activities to non-central planning matters, see Städtisches Anzeigenblatt Frankfurt a.M. on 26 August 1933, p. 378f, ISG.

officials when several days later, on the 24th of August, Todt announced a new route which was to be built – the Munich–Salzburg route – without having consulted with Hafraba experts.[610] This was – after the Frankfurt/Main to Darmstadt route that was finalized on 24. Juni 1933, the second route to be officially defined by the National Socialist regime.[611]

On the same day of the press release, 24 August 1933, Todt had already surveyed the route together with the head of the *Reichsbahn*, Dorpmüller, the president of the railroad's administrative board, von Siemens, as well as Minister of State Esser with a sizable staff of experts and specialists.[612] In connection with the announcement of the new route, Todt officially released a rough sketch of the course to the media. Todt's announcement came like a downright coup that particularly took Munich's local politicians by surprise.

The worried councilman Harbers of Munich, for example, asked Minister of State Esser if Munich's experience with regional planning could be incorporated into the expressway plans.[613] In September of 1933 both Woerner and the media issued route descriptions in greater detail.[614] After founding the *Reichsbahngesellschaft* on 25 August,[615] the main construction management for the expressway Munich–National border (Salzburg)“ took up its work under the leadership of the *Reichsbahn* official Doll, in Munich's Arnulfstraße 32 on 1 September 1933.[616] On September 20 1933, the communities that were affected by the expressway construction were informed of the project during a meeting at the *Reichsbahn* building in Munich.[617] The route's final course was to be determined quickly and at Hitler's urging construction was to begin soon.[618] Delays caused by the question of how the Neubiberg

610 Münchener Neueste Nachrichten, 25 August 1933, Staatsarchiv Munich, Autobahn, München–Salzburg 1933–1934.

611 Völkischer Beobachter on 25 June 1933, p. 2, Edition A, BA, Die Frankfurter Nachrichten on 30 June 1933 front page headlines on beginning of surveying work in June: „Ein Paradies für Autofahrer“, ISG.

612 Münchener Neueste Nachrichten on 25 August 1933, SM, Autobahn, München–Salzburg 1933–1934.

613 Letter on 28 August 1933, SM, Verkehr 84.

614 Bayerische Staatszeitung 6 September 1933, op. cit., Woerner, Anton, op. cit.

615 Business report on the first year 1933, in: Die Autobahn, issue 10, 1934, p. 469.

616 Letter from main construction management to Munich city council on 9 September 1933, SM, Verkehr 80.

617 Meeting report,Die Autobahn, issue 10, 1933, p. 21.

618 Letter from Todt to head of supreme construction, Doll, on 5 October 1933, Bayerisches Staatsarchiv, Munich, Autobahndirektion Südbayern, 248.

Airport was to be tied into to the route, however, led to problems in planning the expressway connection to downtown Munich.

Systematic route negotiations were held with the affected communities and available documents show that invitations for meetings were received by the local councils of Pank, Aising and Pfraundorf on January 2, 1934. The mayors of Obersiegsdorf, Untersiegsdorf and Kirchleiten also received invitations on this day. The time period between November 7, 1933 and January 15, 1934 witnessed gatherings in the communities of Hofolding, Valley, Föching, Pank, Aising, Pfraundorf, Reichersdorf and Irschenberg.[619]

Construction work began with tree-felling work at the beginning November 1933 on a 20 km-long segment in the Hofoldinger Forest. Up to 200 lumberjacks were active at a given time.[620] Work on contract section 2 (km 4,5 to km 12,2) in Unterhaching-Brunnthal – along with five expressway bridges, was contracted out on 15 November 1933. This construction site employed 700 workers at the beginning of December. Forest clearance work and top soil removal in contract section 1 (km 0,5 to km 4,5) was contracted out in December 1933.[621] The circumstance that Hofolding Forest was a state forest and required no clarification of property rights, allowed construction work to begin here early. The situation at the Frankfurt–Darmstadt leg had been similar as many of the first kilometers passed through a city and state forest respectively. In anticipation of the winter construction break, the foundations for the Inn Bridge were completed by December of 1933.[622] When the expressway construction work officially began on 21 March 1934, work was already underway on 44 km of the total length involving a work force of 2700. Five hundred of these laborers had been newly hired in March 1934.[623] At Todt's urging, on the first of June 1934, the main construction management in Munich agreed to produce ready-for-building blueprints for contract sections VI through

619 Bayerisches Staatsarchiv, Munich, Autobahndirektion Südbayern, 0223.

620 Münchner Augsburger Abendzeitung 20 February 1934, SM, Autobahn, München–Salzburg 1933–1934.

621 Ebenda. For construction data see Friedrich Doll: Vom Bau der Reichsautobahn München–Landesgrenze, in: Die Strasse, issue 5, 1935.

622 Münchener Neueste Nachrichten on 24 December 1933, Bayerisches Staatsarchiv, Munich, Akte IHK 117, Reichsautobahnen.

623 Bayerischer Kurier on 21 March 1934, SM, Autobahn, München–Salzburg 1933 bis 1934.

XIV by 1 July 1934.[624] The Lake Seehammer–Siegsdorf route had been sub-divided within these contract sections.

There were several points of contention during the execution of the construction work. To begin with, several farmers could no longer access their fields where construction work had trenched their properties and temporary crossovers had not been prepared. This resulted in numerous petitions and complaints. The extensive damage to the construction site access roads led to countless other complaints.[625] Its high priority for Hitler explains the intense efforts to complete this leg. By the end of 1934, there was construction activity over the entire length of the 100 km leg from Munich to Siegsdorf. A total of 11 750 men worked in two or even three-shift operations.[626]

The expressway's width measured 24 meters. The roadway in each direction had a width of 7.50 meters over two lanes. The median strip had a width of five meters. A two-meter wide shoulder supplemented the lanes. This width, thus, noticeably exceeded the 20.50 meter width of the Frankfurt–Darmstadt leg that had been planned by Hafraba.[627] From the Siegsdorf exit to the national border, the width was reduced from 24 to 17 meters.[628] Here, the shoulders were omitted and the median strip was reduced to two meters. An anticipated reduced level of traffic was put forth to explain the reduction – an assumption that might have been related to the Austria ban for German tourists (see below). Interestingly, the expressway's reduced width starting from Siegsdorf is still in place today.

The Table 6Table 6 above summarizes the inaugurations of the first expressway segments from 1935 to 1936 throughout the German Reich.[629] The table reveals that the Munich–

---

624 Letter from Doll to Todt on 1 June 1934, Bayerisches Staatsarchiv, Munich, Autobahndirektion Südbayern, 248.

625 Ibidem.

626 Friedrich Doll: Vom Bau der Reichsautobahn München–Landesgrenze, in: Die Strasse, issue 5, 1935, p. 157.

627 Uhlfelder: Richtlinien für den technische Anlage von Autobahnen, Hafraba-Mitteilungsblatt 4/1931, p. 5.

628 Doll, op. cit., p. 432.

629 Statistisches Jahrbuch für das Deutsche Reich, 1936, p. 198. Five expressway segments were inaugurated simultaneously to mark the end of the Olympics in Berlin on 17 August 1936. Propagandists were hopeful that the assembled international press corps would cover these events.

Salzburg leg was the highest priority; of the altogether seven segments that were opened to traffic through 24 May 1936, three were on this leg.

A much-reported on large-scale rest stop was built on the Chiemsee's shore to heighten the appeal of traveling on the expressway. Among others, Windisch-Hojnacki researched this rest stop extensively.[630] The rest stop was investigated with a view to its function to encourage drivers to linger and also to provide the conveniences that heightened the enjoyment of travel. Drivers were lured to the rest stop with the prospect of „taking a refreshing dip – directly from the expressway – in the cool waters of the Chiemsee on a hot summer's day".[631] The rest stops would also allow long-distance bus travelers the opportunity to take a break. A third function of the rest stop was to serve as an en-route stop for travelers from Munich on their excursions. This project was thus planned as a large-scale facility.

This extended facility encompassed a hotel, a restaurant and a terrace with seating for 1300 guests. With considerable expenditure and effort, it was built from 1937–1938 on piles in the water. Seifert designed all of the outdoor-facilities.[632] The interior fittings of both the hotel and restaurant were designed in a Bavarian country-home style with handy-crafted objects to display quality German workmanship. The Munich-based carpenter's workshop of Josef Hallhuber provided the fine cabinetry. The large seating area displayed a wall decoration depicting „alpine folk types." by the Garmisch artist Heinrich Bickel. The chandeliers were created in the metal smith workshop of Reinhold Kirsch of Munich. The hotel rooms had

---

630 Windisch-Hojnacki, op. cit., pp. 97–99. Further publications on the rest stops: Angela Schumacher: „Vor uns die endlosen Strassen, vor uns die lockende Ferne..." – „Vom Tanken und Rasten auf Entdeckerfahrten durch deutsche Lande", in: Rainer Stommer and Claudia Gabriele Philipp (eds.): Reichsautobahnen – Pyramiden des Dritten Reiches, Marburg 1982, pp. 77–90, here p. 85s. and: Steininger, Benjamin: Raum-Maschine Reichsautobahn, Berlin, Kulturverlag Kadmos 2005.

631 Hubert Wilm: Architektur und Landschaft – Vor der Eröffnung der Autobahngaststätte am Chiemsee, in: Münchner Neueste Nachrichten on 12 August 1938, SM, München–Salzburg 1936–1939.

632 Wilm: Architektur und Landschaft, in the place cited. For a floorplan of the rest stop facilities see Ralph Johannes and Gerhard Wölki: Die Autobahn und ihre Rastanlagen – Geschichte und Architektur, Petersberg 2005, pp.73–83. Also: Anton Grüb: Reichsautobahn-Raststätte am Chiemsee, in: Die Strasse, 1938, issue 7, p. 217 and Hafen, Peter, op. cit.. Hafen's report includes photos and a floorplan of the facilities. From 1945 to 2004 the facilities were used by the US Army. Post-1945 photos can be found in: John Provan: The Chiemsee Hotels, Verlag Plenk, Berchtesgaden, year not indicated (ca. 1985). My attention was brought to the Provan book by Gerhard Wölki for which I am thankful.

furniture that was painted in „peasant style."[633] The facility was complemented by the swimming pool by the sea. For the grand opening of the restaurant on Saturday, August 27, 1938, visitors came in great numbers from the entire Oberland region and as far away as Salzburg, Tirol, and Berchtesgaden. Many visitors arrived from Munich as well. Reporters who covered this event also noted the presence of foreigners from Italy, Holland, France and England.[634] Altogether 8000 visitors who had arrived in a total of 1800 motor vehicles and 60 buses were counted on this day. They drank more than 5000 cups of coffee. Image 65 shows the rest stop with the parking lot.[635]

Image 65: Rest Stop at Lake Chiem

The martial language that was used by the Bavarian press corps to document all phases of the construction work involuntarily highlighted the poor working conditions at the construction sites. A collection of press-clippings in the City Archive of Munich documents this. The kick-off on March 21, 1934 was described as "a major day of action in the battle of labor." The

633 J. Greiner: Ratshaus am Chiemsee – Musterbeispiel deutscher Wertarbeit, in: SM, München–Salzburg 1936–1939, likely Münchener Zeitung on Saturday, 3 September 1938. Also: Hubert Wilm: Vom Zauber heimischer Bauweisen – Auch der Hotelteil des Rasthauses am Chiemsee vollendet – Kleiner Gang durch das Haus, in: Münchner Neueste Nachrichten on 29 January 1939, SM, München–Salzburg 1936–1939.

634 „3000 cars on the first weekend" – title of the Münchner Neueste Nachrichten on 29 August 1938 on the opening of the Chiemsee rest stop, SM, München–Salzburg 1935.

635 SM, Munich–Salzburg 1936–1939, Völkischer Beobachter on Saturday, 3 Sepember 1938.

miserable morale of the laborers, it should be noted, was in a startling contrast to the glamor that attended the expressway propaganda. A report from the district office of Rosenheim pointed out in April of 1924, that the bearing and morale of the emergency workers at the expressway was very bad. An appraisal from Munich's police directorate in 1935 suggested that the expressway construction sites could be described as „a breeding ground for communism“[636]

21.6 The Orchestration of the Expressway Construction as a Propaganda Campaign

The beginning of the expressway construction was celebrated officially in the Reich on March 21, 1934 with the staging of a propaganda spectacle. For this event, Goebbels timed the construction work to begin on the same day at 22 separate construction segments throughout the Reich.[637] Also that day, he had Hitler's speech at Unterhaching, to mark the start of construction in Bavaria, broadcast over loudspeakers to the assembled workers from all the other sites. Image 66, taken on March 21, 1934 shows unemployed men in Unterhaching who were forcibly recruited to work at the expressway construction sites.[638]

The propaganda-tinged reporting about the expressway drew many farm laborers to the construction sites in the hope of finding work. The goal, however, of providing work for the unemployed could not be completely achieved by the expressway project. In a reaction to this, the State Ministry for the Interior issued a sharply worded letter to the district administrations on 8 March 1934 stating that job centers must under no circumstances accept farm laborers for work at the construction sites.[639] The Großglockner high-alpine roads similarly attracted many farm laborers to the construction sites; they too were turned away in favor of the unemployed who were registered at the job centers.[640]

---

[636] Ulrike Haerendel and Gabriele Krüger: „Groß-München“: Eingemeindung, Verkehr, kommunales Bauen, in: München-Hauptstadt der Bewegung, edited by Stadtmuseum Munich, 2002, pp. 287–293, here p. 287.

[637] Die Autobahn, issue 5, 1934, p. 168. See also letter on 12 February 1934 from Thuringian division of Propaganda Ministry to the President of Thuringia, in: Main Thuringian Archive, Weimar, C234, p. 19.

[638] Franz Ritter von Epp: Ein Werk der Einheit, in: Die Strasse, issue 12, 1935, p. 430.

[639] SM, Verkehr 80. Later a Reichs-wide regulation prohibited farm workers from being placed at the construction sites, see Die Strasse, issue 3, 1934, p. 92.

[640] G. Rigele, op. cit., chapter 4.

Image 66: Forced laborers in Unterhaching at the Beginning of the Expressway Construction

Major propaganda efforts continued unabated over the remaining course of the expressway construction. Todt invited Munich media representatives to report on his travels to the expressway construction sites aboard the airship Graf Zeppelin.[641] On 9 June 1934, a wandering exhibition on *„Die Straße*" (The Road) was opened in Hitler's presence.[642] After being shown in other cites of the German Reich, the exhibition was also shown in the occupied territories after 1938. The *Straßenbauausstellung München 1934 e.V.*, (Association for the Exhibition of Road Construction), whose managing director was Professor Lechner, an architect from Munich, organized this exhibition. It took place in Hall 1 of Munich's exhibition area on the Theresienwiese. As Todt mentioned in a letter to Seifert on Jan 18, 1934, the idea for this exhibition had been personally suggested by Hitler. In this letter, Todt sought out Seiferts participation.[643] The overall purpose of the exhibition was to present in a popular way the important correlations of public transport to business, work, science, regional planning, architecture, transport and administrative organizations. The entrance to the exhibition hall featured eight monumental frescoes depicting the historical development of roads from a

---

641 Münchener Zeitung on 22 May 1934, SM, allgemein 1933–1934.

642 Die Autobahn, issue 8, 1934, p. 288.

643 ADM, 133/56, letter on 18 January 1934.

Germanic dirt tracks to the expressway.[644] At the exhibition's center was the recently begun expressway project under the slogan „clearing the path for motorization." An over-dimensional wall map of the planned expressway network that had been prepared by the artist Zietara of Munich was also on display. Another major focus of the exhibition was the New Automobile section where the German automobile industry presented their most modern vehicles ranging from a compact to an omnibus.[645] In the „road and landscape" section of the exhibition, Seifert presented a model of the expressway at Irschenberg which was detailed with models of the vegetation that he envisioned for this route section.[646] The NS regime's much debated push for modernization is underscored by the exhibition's role in encouraging higher motorization levels.

The exhibition „Die Strasse" was integrated into the 7th International Road Convention that was held in Munich from September 3–9, 1934. This convention, which was inaugurated by Reichsminister Hess and Todt, provided a welcome platform to present the German expressway project to international road construction experts. On September 6, 1934, participants of the convention were given the opportunity to join in an excursion to the constructions sites for the Munich–Salzberg expressway and the alpine crossroad.[647] Several months later, on 18 February 1935, the exhibition „*Die Strasse*" was also shown at Berlin's „International Automobile and Motorcycle Exhibition" in Halle 1.[648]

Numerous speeches were held in Munich to popularize the expressway project. Seifert, lectured on the topic „Trustees of the Landscape – harmonizing the expressways with the landscape" for the Upper Bavarian Society of Architects and Engineers in Munich.[649] Woerner held a speech on the Munich–Salzburg expressway project in February of 1934 before this

---

644 Lang, K. and R. Stommer: Deutsche Künstler – an die Front des Strassenbaues! – A case study of the national socialist art genre „Autobahnmalerei" (expressway art), in: Stommer and Philipp, op. cit., p. 97, which depicts two frescos.

645 Article about the exhibition in Die Autobahn, issue 8, 1934, p. 289. Also see the official exhibition guide.

646 Letter from Seifert to Todt on 12 May 1934, ADM, 133/56.

647 Völkischer Beobachter on 4 September 1934 and 5 September 1934 with speeches, the convention program, and an essay from Seifert, Ausgabe A, BA, see also Die Strasse, issue 2, 1934, pp. 34–40. Hess referred to the monument function of the expessways: It is a „political-historical document of our era by which posterity will judge us."

648 Die Strasse, issue 4, 1935, p. 121s.

649 Münchener Neuste Nachrichten 24 November 1934, SM, Autobahnen allgemein 1933–34.

society's members as well.[650] Todt spoke on "The state of road construction work in the Third Reich" before an audience that filled to capacity the physics auditorium of Munich's Technical University on January 20, 1936.[651]

The incorporation of art in the expressway project is yet another aspect of the propagandist's work. The scope of Todt's efforts to exploit art for his goals is surprisingly extensive. But these efforts were uneven and some of the expressway routes Frankfurt(Main) to Darmstadt, for example, show no evidence of art involvement at all. It has been established by Windisch-Hojnacki and Lang/Stommer that Todt hired numerous artists to document the progression of the construction work throughout the entire expressway network. Some of the expressway paintings were prepared as a series. The artist Fritz created 116 monochrome drawings and another artist named Huber filled up almost 200 folios with water colors of the scenes he observed.[652] Other expressway paintings were created by the Todt-commissioned artist Vollbehr who made his observations on the Munich–Salzburg leg – including a work depicting the Prien Bridge in an anti-rust coating, see Image 67. Vollbehr had been a fellow soldier who served with Todt during WWI where he was already practicing his craft on war scenes. Todt had hired him in 1933 to track the work on the expressway from an artist's view.[653] Later Vollbehr would use the fortifications at the border to France as subject matter for his paintings and also paint scenes from World War II. A passage in his book's introduction reveals that Vollbehr viewed the expressway project as a battle in a war.[654] Of the 40 paintings that are presented in his book as four-color-reproductions, 31 are bright-colored representations of bridges. In a stark contrast to his paintings on the subject of „the battle of labor," these paintings are all but devoid of people.

---

650 Münchner Augsburger Abendzeitung on 20 February 1934, SM, München–Salzburg 1933 bis 1934.

651 Völkischer Beobachter on 21 January 1936, SM, München–Salzburg 1935–36.

652 Windisch-Hojnacki, op. cit., pp. 199–224, Lang and Stommer, op. cit., pp. 91–110.

653 About Vollbehr see: Lang/Stommer, op. cit., p. 97. Windisch-Hojnacki, op. cit., on pp. 217–222 interprets the work of Vollbehr.

654 Ernst Vollbehr: Arbeitsschlacht – Fünf Jahre Malfahrten auf den Bauplätzen der „Strassen Adolf Hitlers", Berlin, 2nd edition, 1938.

Image 67: Vollbehr Painting of the Prien Bridge Bridge during Construction

Windisch-Hojnacki interprets this emptiness in Vollbehr's paintings as an attempt to stress the technical aspects of the bridge constructions. The fact that 50 000 copies of this volume had to be reprinted in 1938 suggests that the book met with an avid readership. Todt's systematic inclusion of so many artists in the expressway project is proof that he understood the project as a major propaganda staging for the Third Reich.

In the presence of several high-ranking guests, Todt opened an exhibition at Rosenheim's council chamber featuring 100 works by K.A. Arnold, an artist from Prien, on May 29; 1935 the various sketches, studies and canvases showed the progression of construction work at several of the sites.[655] In February 1936, artists throughout Germany were invited to submit work for an exhibition on „Adolf Hitler's roads“ to be held in September of that year. In a bid to inspire Munich artists, and to prepare them for the competition, three buses were provided in May for excursions to the Mangfall Bridge, Siegsdorf and Bergen, as well as to Inzell and Mauthäusl for the completed section of the alpine crossroad. The exhibition put on display a selection of 500 of the 1200 submitted works.[656]

---

655 Münchener Zeitung on 29 April 1935, SM, München–Salzburg 1935–36.

656 Völkischer Beobachter 24 May 1936, SM, München–Salzburg 1936–39, Claudia Windisch-Hojnacki, op. cit., p. 200.

Monumental expressway structures – especially bridges[657] – were embedded within the concept of landscape aesthetics as a symbol for eternity to heighten the prestige of the Third Reich. It appears that monumentality was pursued without regard to costs. Indeed, landscape aesthetics and monumentality were goals that had priority over economic efficiency. The large-scale bridge constructions over the Mangfall and Inn Rivers were especially suitable for the propaganda efforts underway. They were widely reported on by Munich's media and celebrated as first-rate feats of German engineering. Image 68 shows the Mangfall Bridge which juts out from its rural surroundings like a symbol of a new era in the winter of 1934/1935.[658] The expressionist style of the photo's composition was unusual within the body of expressway propaganda. As Windisch-Hojnacki was able to demonstrate based on the expressway paintings, the Mangfall Bridge stood at the center of a special Nazi propaganda campaign.[659] Apart from Vollbehr, such artists as Merkers, Geigenberger and Fritz also produced artwork on this subject.

The propaganda directors from Munich's section of the Goebbel Ministry came together with Todt for the inaugurations of the individual construction sections. The opening celebrations were orchestrated as spectacular propaganda events – Schütz und Gruber were able to identify at least five such opening ceremonies for the Munich–Salzburg leg between 1935 and 1939.[660] The opening of the Weyarn–Samerberg expressway section, for example, was concluded with a viewing of K.A. Arnold's exhibition entitled „Two Years of the Reich's Expressways“ for the high-ranking guests at Herrenchiemsee Castle on May 24, 1936.[661]

---

657 See issues 8 and 23 of Todt's journal Die Strasse, Jahrgang 1935, with a special focus on expressway-bridges, also later issues. Further: Claudia Windisch-Hojnacki, op. cit., Chapter 7, and R. Stommer, Triumph der Technik – Autobahnbrücken zwischen Ingenieuraufgabe und Kulturdenkmal, in: Stommer, op. cit., pp. 49–76. Also on expressway bridges: Bernd Kreuzer: Tempo 130, op. cit., pp. 74–85. In the US, special attention was paid to expressway bridges as well. For info on the Jugendstil and Art Deco elements on bridge constructions of the parkways in the US see Bruce Radde: The Merrit Parkway, Yale University Press 1996.

658 SM, München–Salzburg 1933–1934.

659 Claudia Windisch-Hojnacki, op. cit., pp. 212–213.

660 Op. cit., p. 59.

661 Völkischer Beobachter on 25 May 1936, SM, München–Salzburg 1936–39.

Image 68: Picture of the Mangfall Bridge that was used for Propaganda Purposes

It is interesting to see how the issue of expressway tolls evolved. In the legal act to establish the *Reichsautobahngesellschaft* on June 27, 1933, paragraph §7 states that road tolls can be levied.[662] Similar to such projects in Italy, the expressways were initially viewed as analogous to the railroads and the expressway-access roads were referred to as „expressway stations" (or "*Autobahnhöfe*"). Comparable to the English turnpike system of the 18th century, these would be equipped with a barrier at which a road fee had to be disbursed for each car before passing.[663] Even up until 1934, Todt talked about road fees that would be collected on the Munich–Salzburg route when facing Munich's representatives.[664] Presumably he took this position to enable the expressway project. The simple national comrades („*Volksgenossen*") viewed the project with skepticism because the high costs of driving (the acquisition of a car, high gasoline prices) would limit this activity to those of the upper income levels, as well as party officials.[665] But once the route to Holzkirchen was inaugurated amidst great jubilation, there was no longer any mention of a road toll. A refinancing of the expressways, however modest, was to be achieved via an increased tax on gasoline as this was considered to be less intrusive than a toll fee for every journey on the expressway.[666] With this, the expressways became a popular „*Volksprodukt*" and went on to take their place among the other folk products of the Nazi consumer society that have been researched by Wolfgang König.[667] What is remarkable is that the toll-free operation of the expressways was continued into the later Federal Republic of Germany.

---

662 This law is reprinted in: Die Autobahn, issue 7, 1933, p. 1s.

663 „Wie die Reichsautobahnhöfe aussehen", in: Der Führer, 1 March 1934, Stadtarchiv Karlsruhe, 1/H-1596.

664 Münchener Zeitung on 16 January 1934, SM, München–Salzburg 1933–34.

665 Dorothee Hochstetter: Motorisierung und Volksgemeinschaft, München 2005, p. 159. An official proclamation about the renouncing of toll fees was not made until the beginning of 1937 on the occasion of the automobile exhibition, see Todt, Fritz: Reichsautobahn und Motorisierung, in: Die Strasse, issue 4, 1937, p. 92.

666 Eduard Schönleben, Department head in Todt's administration, writes about the tax increases to finance the expressways in his contribution: Omnibus und Güterkraftverkehr als Kostenträger der Reichsautobahn, in: Die Strasse, issue 13, 1937, pp. 363–365. Kaftan indicates the substantial sum of 1.9 billion Reich-marks that were transferred to the Reichsautobahngesellschaft from custom and tax revenues through 1945, op. cit. p. 162.

667 Wolfgang König: Volkswagen, Volksempfänger, Volksgemeinschaft – Vom Scheitern der nationalsozialistischen Konsumgesellschaft, Paderborn 2004.

22 Experiences with the New Routes and their Effect

It would be surprising if the experiences made by driving on the expressways, the country roads and the railroads were comparable. Where there qualitative differences to a ride experienced in a car, a bus or a motorbike? There are not many sources which contribute to answering this question. In a reportage about a bus ride from Kassel to Frankfurt on the expressway, tourist highlights such as Burgen and the intensity of the experiences were emphasized: how passengers witnessed the beauty of the landscape. The author had the impression that the landscape addressed the travelers more directly and powerfully than a train ride because of its closer proximity. The traveler sees the landscape hasten towards him or her.[668] Car drivers also remarked on how forcefully landscape is experienced compared to a train ride. The panorama view of the passenger in the train compartment, as described by Schivelbusch, is replaced by the traveler's gaze at the onrushing landscape through the windshield.[669]

The liberal journalist Walter Dirks, who would become known as the publisher of the *Frankfurter Hefte*, reflected on a car ride experienced on the expressway from Frankfurt to Göttingen via Kassel in a reportage for the *Frankfurter Zeitung*. He was traveling on the occasion of both the opening of the route Frankfurt a. M.–Kassel and the ceremonial act for the completion of the first 3000 kilometers stretch of expressway. He described his experience to the effect that the relationship of car to lane had been reversed. It is not the car that drives on the lane, on the contrary, it is the lane that moves swiftly and smoothly, so as to virtually suck the car in. But let us hear him here with the reportage itself:[670]

668 Rolf Mengel: Die schnellen Riesen der Reichsautobahn, in: Kurhessische Landeszeitung, 14. January 1939, Stadtarchiv Kassel.

669 Schivelbusch, op. cit., Eisenbahnreise, 2000. The significance of the gaze through the windshield for artistic perception is described by Jason Weems: Theorizing the Windshield: Vision as the Edge of Mobility, T2M Conference, University of Michigan-Dearborn 2004.

670 Walter Dirks, in: Frankfurter Zeitung, Reichsausgabe, 11.12.1938, Nr. 631, p. 5. I owe this source to my colleague Eberhard Schütz. The complete text of the reportage is published under the title „Das Dreieck auf den Autobahnen – Impressionen einer Fahrt Frankfurt–Berlin–München–Frankfurt" in: Walter Dirks: Feuilletons im Nationalsozialismus, Gesammelte Schriften, volume 3, edited by Fritz Boll, Ulrich Bröckling and Karl Prümm, Zurich 1990, pp. 136–152.

The triangle on the expressway

When reporting about this drive, then its beginning can be skipped with just a few words, which passed over an all too well-known route. But then this route brought with it straightaway a surprise: it was shorter than usual. We were surprised by the sign "Ende der Autobahn" at Reiskirchen, when we imagined ourselves to be shortly behind Nauheim. On that hazy November day the gap (meanwhile closed on December 4th) between Reiskirchen and Romrod was still open. The half an hour on the hilly Reichstraße, neatly planted with trees, was a pleasant change, despite hens and confusing curves. It was still early morning, before eight o'clock, the streets were empty, and whoever drove or walked on them still had the morning quiet in him, including us ourselves.

Past the small Upper Hessian and Kurhessian towns, past Alsfeld and Homberg, whose silhouettes, slightly veiled in the distance, stood at the horizon, light grey behind the brown and green of the fields and winter grains; the expressway still seems to run past these small towns without connection: there doesn't yet seem to be a sign that here something substantial has changed. When the working camps have been demolished, the construction machines have departed and the wounds on earth have become overgrown with fresh grass, the double strip will run thither smoothly and just for itself, with a momentum which stands in no relation to the well-ordered, resting world, of which those small towns are important centers. That is how it seems to be from the expressway on which we, in midst of Hessian woods, still imagine ourselves to be in the center of events, within the power current of large cities. In the workshops, living rooms and inns of Alsfeld and Homberg the very opposite may be felt: Here, in work, fight, love and hate, real life flows, and five kilometers out there, over the fields of the farmer N. and the woods of the squire M., a road runs past on which strange, invisible people drive from Frankfurt to Berlin, what's that to us? That is the way it is today. In ten years Alsfeld will no longer lie so disconnected near the expressway. Then one will be able to read in the city's image that it takes part in the traffic lanes.

The absorbing lane:

Was it in the mountains, over which the expressway boldly strides, or was it as early as Rimberg where the lane, with one single sweeping curve circuits a whole mountain, almost attaining the height of a summit plateau. On those steep ascensions is where the "lane" unveils its true nature. It is not a road on which we move forward through brisk activity, past that tree and then that house, around this obstacle, then stopping before that other one, braking, accelerating, shifting, clutching, honking. On the road it is we who drive; the motor only lends us the "raw" power to do so. On the lane, however, we do not seem to be active anymore; almost imperceptibly and barely conscious do we regulate and control the car, which we have put into motion and that now, without assistance from us, runs on the strip of lane. Yes, we even feel that the expression "runs on" is not really apt, because we no longer imagine the running wheels but rather the gliding car. So very passive are we and so strong is the impression of the sweeping lane that finally the relation seems to inverse: It is the lane which is active; it moves swiftly and smoothly without friction and force towards us and draws the car irresistibly into itself. It draws us through the terrain, sometimes it even draws us through the landscape. Because there are places – the place at Rimberg is one of them, the most striking could be the spot behind Kirchheim below Teck, where the lane leading from Stuttgart to Ulm, on a long, high, curved bridge, bores into the precipices of the Alb – where the swaying lane not only overpowers the shape of the terrain, but becomes inextricably linked with an essential part of the landscape. It draws us into the valley, over a mountain slope, around a summit, so that we experience these shapes and forms first hand and as new. Speed is required for this impression to arise, but most important is the movement's smoothness and lack of friction. It is as if our body itself, safe within the sleek metallic capsule, were to track and scan the forms of this landscape by swiftly and speedily gliding on the curves of the expressway. On the roads there is nothing like it, even on the most scenic mountain pass road: only the eyes can absorb here, purely and without hindrance, the landscape; otherwise a thousand prior trifles of traffic and the road surface stand between the body and the landscape.

About Harz:

We glided down the slope of the Knüll, at our right lay *the Fulda, deeply incised into the valley. Kassel remained a maze of buildings on our left, the dividing woods of Habicht with the pointed obelisk of Hercules aloft, and the wide white castle above the city, from which the street descends dead straight into the core of the settlement: all that abided, veiled by mist. Very soon new woodlands enveloped us; and as the wall of trees opened to our right and left, for two or three seconds in the depth, nestling in the brown hills, Hannoversch-Münden appeared, or rather its fleeting image: a few towers, half-timbered houses, smoke from the chimneys. Then the lane sank toward the Werra and crossed it on a bridge, which we remembered from a striking photograph; it is one of the greatest constructions of the expressway. But here we learned that expressway bridges look much more impressive from below than from the lane itself. The bridge is seamlessly inserted into the lane, it wills to, and should, overcome the valley, make it vanish. Those who are not informed and drive slowly and sit straight up, hardly notice that they are crossing a river, no booming masses of metal and no towering arches remind them of the same.*

A quarter of an hour later, in Göttingen, the expressway had ended for the time being. We saw ourselves before the only large gap in the expressway network in the triangle Frankfurt–Berlin–Munich that will remain open to the end of this year.

So far Dirk's text.

The *Wiesbadener Autoverkehrsgesellschaft* (Wiesbaden Society for Transport by Car), which was specialized in package tours and day trips for spa guests and citizens of Wiesbaden, painted a rather more negative picture. According to their experiences on the expressway Frankfurt a. M.–Heidelberg, long trips on the expressway had a monotonous and tiring effect, which was the reason why they partly favored slow journeys on picturesque country roads through scenic regions distinguished by outstanding buildings.[671]

In a reportage for the *Frankfurter Zeitung* from 22 May 1935, on the occasion of the opening of the route Frankfurt a. M.–Darmstadt, the reporter rendered his impressions of the journey with the streamlined bus of the *Reichsbahn*: The passenger could see through the slightly inclined windows, that recalled skylights, how the landscape in its breadth glided past. The strangest thing about this travel experience, of almost flying but still being tied to the road, is the smoothness, the lack of friction of the traffic lane.[672]

The effect the expressways had on the supply of food for larger cities was examined with a classic geo-economic approach by Carl Lüer, President of the Chamber of Industry and

---

[671] Verkehrstechnik, issue 4, 1936, p. 97.

[672] AHR.

Commerce for the economic region of Rhine-Main, in an article published in *Die Straße*.[673] He differentiated thus, following von Thünen's location theory from 1826, the various zones of production of different kinds of food that are located ring-like around a city. Apart from the supply of beef from the *Frankfurter Schlachthof* (slaughterhouse of Frankfurt), Lüer also addressed Frankfurt's milk supply. He demonstrated how the motorization of road traffic had produced a concentration of fresh milk suppliers through the *Milchsversorgungsverband Rhein-Main* (Milk Supply Association Rhine-Main). The dairies located to the north of Frankfurt, purified the region's milk of their region would, via of the expressway Frankfurt–Gießen, play a crucial role in supplying Frankfurt.

Lüer also emphasized the significance of the expressways for the numerous creameries of the German alpine region, whose distribution area could now be extended to the north into the economic region of Rhine-Main. Before there was motorized transport, it had been necessary to transport fresh milk via the railroad in a rapid and well-synchronized supply chain where only milk producers in the vicinity of railroads could be considered.[674]

In the NS-era ideas were again taken up – and imbued with NS-specific ideology – from settlements debates in the 1920s, which focused on the dispersal of inner cities and the formation of garden cities at the peripheries. In the second issue of the *Rhein-Mainischen Wirtschaftszeitung* (Business Magazine Rhine-Main) of June 1934, the settlement concept of the NS-movement was introduced by three authors: Feder (*Reichskommissar für das Siedlungswerk* – Reichkommisar for estate development), Ludowici (*Leiter des Heinstätten-amtes der DAF* – Head of the Homestead Department of the DAF) and the alderman Niemeyer.[675] Whereas Feder and Ludowici unveiled general principles for the settlements, Niemeyer explicitly linked the concept of the settlements with the expressway as a transport mode. His reflections on time-distance research led to the comparison of the 1½ hour-access

---

[673] Lüer, Carl: Neue Möglichkeiten der Lebensmittelversorgung durch den Kraftverkehr mit besonderer Berücksichtigung der Reichsautobahn, in: Die Strasse, 1935, issue 9, pp. 334–339.

[674] Schmitt: Die Milchversorgung der Großstädte, in: Der Güterumschlag, Sonderausgabe der Zeitschrift des Vereins Deutscher Ingenieure, Berlin, 1926, pp. 235–237. Here milk supply is discussed, taking the example of Mannheim, Hamilton shows how the von Thünen rings around Chicago in 1930 were used to regulate the milk supply of the city and to grant milk farmers privileges for supplying the city, see Shane Hamilton*:* Trucking Country, Princeton UP 2008, pp. 26–28.

[675] ISG.

radius of both expressways and railroads in the wider area of Frankfurt a. M., whereby he emphasized the substantially larger area which was covered by the expressway.

The *Gau-Heimstättenamt der Deutschen Arbeitsfront* (Gau Homestead Department of the German Labor Front) (DAF) could integrate the expressways' modern transport concept with their expansive settlement plans. This case is remarkable since here two independently planning organizations, DAF and *Reichsautobahngesellschaft*, could subsequently coordinate their work. The *Frankfurter Zeitung* brought the topic up again on 24 January 1937 with the article „*Dezentralisierungswirkung der Autobahn*" ("Expressway's Effect of Decentralization"), measured through the development zones of 1½ hours for working commuters.[676] It reported that the newly-built settlements along the expressway that had been erected by the *Gau-Heimstättenamt* within the development zone could be linked to Frankfurt via omnibus connections. With the NS-state's typical mix of backwardness and modernity, it explained that the settlements should be erected in the proximity of the "food basis of rural communities" – a thesis put forward by Dr Ludowici in his lecture at the *Weltkongress für Freizeit und Erholung* (World Congress for Leisure and Recreation) in Hamburg. In anticipation of the suburb agglomerations of large cities, the settlements were linked to the concept of "landscape areas or city landscapes".[677] Here a "healthy effect" of the expressway was seen.

The generous settlement program along the expressways should, in the following years, demarcate the newly developed landscape areas for new settlements or resettlements from those parts of the old town which were to be cleared. With the support of Gauleiter Sprenger, the new "*Erbhöfedorf*" was erected in Ried in Hesse, between Lorsch and Bürstadt; its estate had been won from fallow marsh and bog land through drainage and dewatering of agricultural land through forced labor. The proximity of the expressway had eased the decision to set up the village on this site.[678] Riedrode was "consecrated" on 10 July 1935, within the context of the *Gau-Tages* (Gau Day) Hessen-Nassau from the *Reichsbauernführer* (Reich Farmer's Leader).[679]

---

676 AHR.

677 Neueste Zeitung Frankfurt, 4.8.1936, AHR.

678 Zibell, op. cit., p. 316f, Frankfurter Zeitung on 24 Januar 1937, AHR.

679 NS-Briefe – Schulungsblätter der NSDAP im Rhein-Main-Gebiet, December 1936, p. 438, 447.

23 Gas Stations on the Expressway

The automobile, as a technical artifact, could only establish itself in society because it became part of a cohesive transport system set up as a comprehensive network in society. Automobile factories, car dealers, roads, legal regulation of transportation and traffic, police training, garages, gas stations, rescue services and emergency facilities, insurances, logbooks, road maps and rest stops all constitute this system.[680] In the following the development and construction of gas stations in the expressway network will be described.

With the development of a network of expressway routes, gradually new gas stations were built. In the location where the country road from Frankfurt a. M. to Limburg crosses – at a roundabout – the beltway Frankfurt a. M.–Wiesbaden, a "gas station adapted to the streamlining of modern transportation" was built. It was considered the largest gas station in Germany and pointed the way for the further development of gas stations on long-distance roads.[681] The first expressway gas station, designed by the architect Bembé, began operating in 1937 at the exit to Darmstadt, cf. Image 69. It had a dynamic sweeping form and a winglike roof.[682] It thus adopted the style of the modern, rounded architecture of the twenties that the architect Luckhardt had used to design buildings at Alexander Platz in Berlin.[683] The gas station Lorsch was built following the same style.[684]

---

680 James Flink describes this for the USA: America Adopts the Automobile, 1895–1910, Cambridge, Mass., 1970.

681 Frankfurter Wochenschau 18 September 1937, AHR.

682 Ralph, Johannes and Gerhard Wölki: Die Autobahn und ihre Rastanlagen – Geschichte und Architektur, Petersberg 2005, p. 17.

683 Verkehrstechnik, issue 11,1929, p. 175, Neueste Zeitung, 28 April 1937, AHR.

684 Operational facilities of the Reichsautobahn, in: Die Strasse, issue 11, 1937, p. 303s.

Image 69: Modern design of the Darmstadt gas station[685]

This modern design of gas stations is also remarkable because in later construction stages the gas stations and rest stops were erected in a "folkloric" style, as postulated by NS-planners, and were adapted to the respective landscapes, such as the one designed by the architect Bonatz, among others.[686]

685 Ralph, Johannes and Gerhard Wölki, op. cit, p. 17.

686 Ralph, Johannes and Gerhard Wölki, op. cit., p. 19. See also Angela Schumacher: „Vor uns die endlosen Strassen, vor uns die lockende Ferne...“ – „ Vom Tanken und Rasten auf Entdeckerfahrten durch deutsche Lande“, in: Rainer Stommer and Claudia Gabriele Philipp (eds.): Reichsautobahnen – Pyramiden des Dritten Reiches, Marburg 1982, pp. 77–90.

24 Development of Traffic

After the opening of the expressway section Frankfurt a. M.–Darmstadt, preparations were made for traffic censuses. As early as 3 June 1935, the inspector general for German road research suggested to the *Landesplanungsverband Rhein-Main* (Regional Planning Association Rhine-Main) in Frankfurt a. M. to carry out a traffic census in cooperation with the main construction office in Frankfurt.[687] This was carried out from 27 May until 2 June 1935 at the feeder road Mörfelden–Langen in both directions and was followed by numerous empirical investigations about traffic on the expressways and a study on the shift of traffic from country roads to the expressway. From 1–7 July 1935, counts were arranged of all vehicles on the highway Darmstadt in both directions. The same took place from 29 July until 4 August 1935 at the feeder road Morfelder–Langen in both directions. Similar counts also took place near Mannheim in September 1935 on the Reich road Darmstadt–Mannheim so as to have comparison data for the development of traffic after the expressway opening on 13 October 1935. The Gezuvor compared the results of its census from 20 August to 27 August 1934 on the road Frankfurt a. M.–Darmstadt to traffic on the same section from 20 July 20 to 5 August 1935 and came up with the following data:

1934: 1771 automobiles, 773 motorbikes and 620 trucks; together 3164 vehicles.
1935: 1240 automobiles, 576 motorbikes and 539 trucks; together 2355 vehicles.

The numbers demonstrate a decrease in traffic on the road Frankfurt a. M.–Darmstadt and render empirical proof of the shifting of traffic to the expressways. The numerous extensive traffic censuses were summed up in an article in the magazine *Die Strasse*, issue 22, 1935. The traffic on the road from Frankfurt a. M. to Darmstadt had decreased by 24.4 percent after the expressway opened. To offer a graphic documentation of this statistic, in the year 1935 a colorful juxtaposition of country road and expressway was made, cf. Image 70.

[687] HHSTAW, Akte 495 Verkehrszählungen auf Reichsautobahnen.

The traffic on the road Darmstadt–Heidelberg had, after the opening of the parallel expressway route, even diminished by 49.2 percent, and on the route Mannheim–Heidelberg by as much as 63 percent.

In issue 22/1935 of the journal *Die Strasse* the gain in time which resulted from using the expressway route Frankfurt–Mannheim–Heidelberg was calculated. The calculations were based on a driving speed of 30 km/h in towns, 60 km/h on the roads and 80 km/h on the expressway.

(1) The route Heidelberg–Mannheim:
Within a total distance of 20 km, there were three villages along the country road. Four kilometers of this stretch passed through these. The expressway was 15 km long and free of crossings – this constituted a 14 minute gain in time.

(2) The route Frankfurt a. M.–Mannheim:
There were 21 towns and villages along this route with a total of 180 mapped crossings and 40.5 kilometers of road passing through them. The expressway offered 70 kilometers free of intersections. Nine kilometers of country road also had to be traveled making a total time saving of 65 minutes.

(3) The route Frankfurt a. M.–Heidelberg:
The standard road along a length of 87.2 km passed 25 localities with 35 km of passage road. In contrast, the expressway was 79.75 km long. Added to this were 8 km Reich road which still had to be driven. The total time gain was 55 minutes.

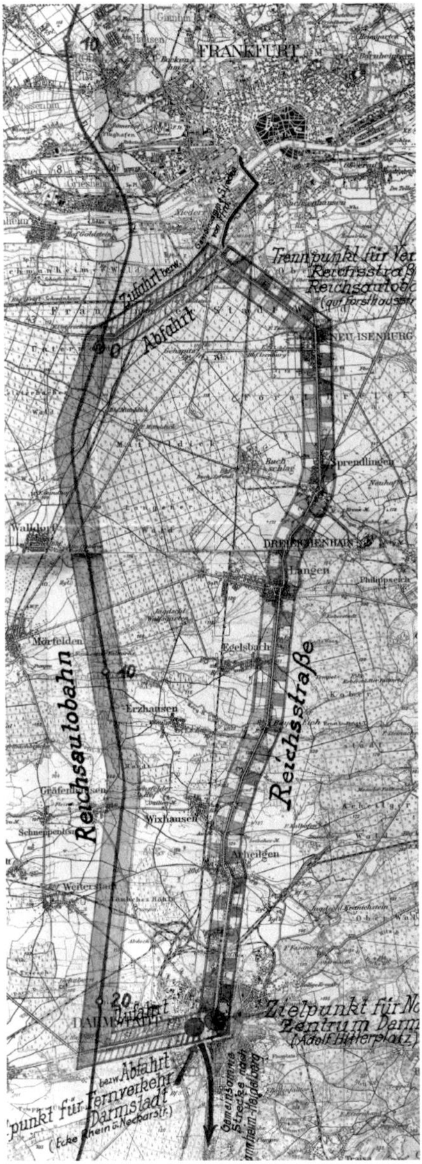

Image 70: The road map Frankfurt–Darmstadt with expressway (solid line) and country road (dotted line)[688]

688 Ibidem.

The data for the excursion traffic at Easter and Pentecost in the years 1936, 1937 and 1938 were gathered from the entire expressway network, because these traffic peaks were particularly apt for propaganda use, which otherwise often had to contend with rather empty expressways. Rudolf Hoffmann (a civil servant in Todt's administration) was responsible for these censuses. In 1938 at Pentecost he had the traffic surveyed at 29 points on expressways across the entire Reich, in addition to 12 counting points on main country roads.[689] The traffic census took place from Saturday before Pentecost at 0 a.m. until Tuesday after Pentecost 12 p.m. Vehicles in both directions were counted. The number of automobiles recorded on the expressways varied from a maximum of 55 000 near Cologne to a minimum of 3000 near Königsberg in Eastern Prussia; the average was 12 000 vehicles. The values measured in the sections of Hesse and Baden were, as a rule, clearly above the average (see Table 10).

Kassel–Hedemünden	7815
Bad Nauheim–Bad Homburg	12411
Darmstadt–Lorsch	24566
Mannheim–Heidelberg	18125
Bruchsal–Karlsruhe	19242

Table 10: Traffic on Pentecost 1938 (both directions)

Excursion traffic on Whitsunday in 1938 was registered at a location between Ramersdorf and Hofolding and provided the data for Table 11Table 11.

Location	Motorcycles	Automobiles	Buses	Trucks	Total
between Ramersdorf and Hofolding	13,217	29,984	835	1,035	45,071
between Bergen and Siegsdorf	9,172	16,445	676	344	26,637

Table 11: Traffic on Expressway München–Salzburg on Whitsunday

689 Hoffmann, Rudolf: Die Entwicklung des Kraftverkehrs auf Reichsautobahnen und Reichsstrassen, in: Forschungsgesellschaft für das Strassenwesen (ed.): Strassenbau-Jahrbuch 1937–1938, Berlin, Verlag Volk und Reich, 1938, pp. 29–42, Hoffmann, Rudolf: Der Kraftverkehr auf deutschen Strassen zu Pfingsten 1938, in: Die Strasse, issue 12, 1938, pp. 388–392, Hoffmann, Rudolf: Der Kraftverkehr auf deutschen Strassen zu Pfingsten 1939, in: Die Strasse, issue 11, 1939, pp. 377–380, Hoffmann, Rudolf: Der Kraftverkehr auf deutschen Straßen Ostern 1939, in: Die Strasse, issue 9, 1939, pp. 283–289.

The 45 071 vehicles that were registered at this point mark the second highest traffic level of the altogether 27 census locations after the Cologne–Düsseldorf leg. When extrapolating the number of counted vehicles to the people transported, a total number of 100 000 vehicle occupants were recorded at Ramersdorf. Assuming that each vehicle made one journey on Whitsunday, it can be estimated that 50 000 people traveled on the expressways from Munich for this holiday. This number corresponds to 6% of Munich's population and is considered to be quite significant. Similar data for Whitsunday traffic in 1939 was released by Hoffmann.[690]

In general, Sunday excursion traffic far surpassed weekday traffic. The censuses that were carried out at the Ramersdorf location on the Chiemsee expressway showed peaks throughout the month of June 1937 that corresponded to levels 100% above the averages for the six workdays.[691]

For the most part, expressway traffic was so sparse that the traffic peaks on Easter and Whitsunday were seized upon with great pride by the propagandists to legitimize the expressway project. Even Todt, the head of the expressway project, used this data to embellish his reports. The progress reports on the expressway project that were delivered at the Nazi party conventions at Nuremburg – the party's main stage – both in 1937 and the following year, referred to the census data taken on this religious holiday; this behavior could be deemed odd for such a brutal and atheistic regime.[692]

When judging the development of tourism in Bavaria in the 1930s, the construction of the Munich–Salzburg expressway was only one of several contributing factors. The ban – at the threat of a substantial fine that the German government imposed on travel to Austria in 1933 –

---

690 Hoffmann, Rudolf: Der Kraftverkehr auf deutschen Strassen zu Pfingsten 1939, in: Die Strasse, issue 11, 1939, pp. 377–380.

691 For this data see Hoffmann, Rudolf: Die Entwicklung des Kraftverkehrs auf Reichsautobahnen und Reichsstrassen, in: Forschungsgesellschaft für das Strassenwesen (Publisher): Strassenbau-Jahrbuch 1937–1938, Berlin, Verlag Volk und Reich, 1938, pp. 29–42, here p. 32s. Weekdays were assigned using Microsoft's Excel calendar mode for this data. Hoffmann reports that the census activity at Ramersdorf was continued without interruption because of a „special order", p. 31.

692 Fritz Todt: Bau und Verkehr auf den Strassen Adolf Hitlers, in: Die Strasse, issue 18, 1937, pp. 518–520, and Fritz Todt: Der Generalinspekteur für das deutsche Strassenwesen berichtet auf dem Parteitag Großdeutschlands über die Strassen des Führers als Schrittmacher der deutschen Bauwirtschaft, in: Die Strasse, Heft 19, 1938, pp. 600–603.

was crucial for the growth of tourism in Bavaria. This penalty was in place from June 1933 to August 1936 (see below). As a result of this policy, the number of German tourists traveling to Austria fell dramatically and tourism to Bavaria, presumably, increased. Indeed, Bad Reichenhall attributed its increased tourist numbers to this policy.

While excellent time series are available on overnight stays in the small tourist villages of Bavaria, it is not possible to use these as documentary evidence for the additionally generated tourist numbers due to a multitude of other contributing factors. The following table provides data for some of the Bavarian resort towns for the summer half-years (20/21 March–22 September) from 1934 to 1939.[693]

The tables indicate either a slight reduction only, or no reduction at all of overnight stays in 1937 which may be attributable to the lifting of the tourist penalty fees for visits to Austria in 1936. Fewer overnight stays were recorded for 1938 and most likely resulted from political and military tensions during the Czechoslovakia crisis. The beginning of the war on September 1, 1939, led to a complete cessation of tourism and is necessarily also reflected in the tourist decreases for 1939. The resort town of Tegernsee registered an increase in 1936 as a direct result of the opening of the expressway exit in July of 1935. Whether the 10.6% rise in overnight stays for Priem am Cheimsee in 1936 was a consequence of the opening of the Rosenheim exit in May of 1936 is unclear because the year 1935 had already seen a marked increase (+38,5%) of visitors. The data for Reit im Winkl showed an increase of 55,1% in 1936 and might be attributable to improved accessibility via the Rosenheim exit. Generally, though, Alfons Woerner's 1933 prediction that the greater accessibility of the resort areas to the east of Munich would strengthen tourism could not be confirmed with this data.[694]

693 Statistische Jahrbücher für das Deutsche Reich 1936 bis 1939/1940.

694 Woerner, op. cit. p. 12.

	SHY1934	SHY 1935	SHY 1936	SHY 1937	SHY 1938	SHY 1939
Number of overnight stays						
Bad Reichenhall	453314	506271	606773	617457	595457	562699
Bad Tölz	271563	288652	277695	275468	269213	233593
Bayrischzell	72849	87872	92724	107997	86525	85048
Berchtesgaden (Land)	947007	902421	979646	1038407	780893	674185
Füssen	144782	158676	156874	182815	143379	141730
Garmisch-Patenkirchen	943765	707075	713135	811292	594448	534453
Mittenwald	292022	254190	312965	406702	323639	269154
Prien am Chiemsee	60645	83979	81444	90048	no data	no data
Reit im Winkl	37276	66616	66662	103423	84762	69655
Ruhpolding	144412	161129	162372	189184	190480	189548
Schliersee	163948	137089	172930	188252	130588	130469
Tegernsee	216054	205069	207396	218651	181409	104562
Change in %						
Bad Reichenhall		11,7	19,9	1,8	-3,6	-5,5
Bad Tölz		6,3	-3,8	-0,8	-2,3	-13,2
Bayrischzell		20,6	5,5	16,5	-19,9	-1,7
Berchtesgaden (Land)		-4,7	8,6	6,0	-24,8	-13,7
Füssen		9,6	-1,1	16,5	-21,6	-1,2
Garmisch-Patenkirchen		-25,1	0,9	13,8	-26,7	-10,1
Mittenwald		-13,0	23,1	30,0	-20,4	-16,8
Prien am Chiemsee		38,5	-3,0	10,6	no data	no data
Reit im Winkl		78,7	0,1	55,1	-18,0	-17,8
Ruhpolding		11,6	0,8	16,5	0,7	-0,5
Schliersee		-16,4	26,1	8,9	-30,6	-0,1
Tegernsee		-5,1	1,1	5,4	-17,0	-42,4

Table 12: Overnight Stays in Bavaria

The easy accessibility of the resort areas owing to the new expressways led to unexpected congestion on the smaller roads that led to the destinations proper. This resulted in the need to increase the capacity of these roads. Tegernsee's mayor complained about the congestion to the Bavarian regional government in 1939.[695] Car traffic prevented tourists from promenading along the lakefront and attempts to broaden the road at the expense of Nazi Fuehrer Himmler's property of all places, resulted in a major dispute. At Todt's intervention these kinds of operations were forbidden.[696]

---

695 Letter on 14 April 1939, Bundesarchiv, Akte NS26/1187.

696 Letter from Todt to the Bavarian government on 1 March 1939, in the place cited.

Although the Munich expressway had been built primarily to carry the excursion traffic to Tegernsee and Chiemsee, it was officially designated the Munich–Salzburg expressway to accentuate its long-distance character. Ironically it was not even possible for German tourists to visit Salzburg in Austria because early in 1933 the Nazi regime had provoked a major conflict with Austria and sharply restricted travel to this country. A fine of 1000 marks was payable by each German tourist who left Germany via Austria.[697] Travel to Austria collapsed as a result of this policy. The proportion of Germans among the tourists who visited Salzburg fell from 52% in 1930/31 to 6% in 1934/35.[698] At short notice, thus, in the summer of 1933 the German travel bureaus had to come up with alternatives for the destinations in Austria. Package tours including hotel accommodation and transportation either by train or bus were redirected to the resort town of Ruhpolding which soon became a popular excursion destination.[699] To avoid being reminded of the embarrassing policies that prevented German tourists from visiting Austria, the expressway designation Munich–Salzburg was replaced by the veiled term „Munich – national border" in such journals as „*Die Autobahn*" and „*Die Straße*" from 1933–1936. Munich's media services used this term as well.[700]

---

697 Reichsgesetzblatt, I, 1933, p. 311. My special thanks to my collegue Hasso Spode for this information.

698 Hellmuth, Th. and E. Hiebl: Tourismusindustrie – Organisation und quantitative Entwicklung des Salzburger Fremdenverkehrs 1860–1938, in: Haas, Robert et al.: Weltbühne und Naturkulisse – Zwei Jahrhunderte Salzburg-Tourismus, Salzburg 1994, p. 95.

699 Roßmann, R.: Mythos Bayern, München 2002, p. 83, and Gabriele Holzer: Verfreundete Nachbarn Österreich – Deutschland, Wien 1995, p. 67.

700 The journal Die Strasse reported about the route "Munich – national border" 18 times between 1935 and 1936, also:SM, Autobahn.

25 Bus Rides on the *Autobahn*

When in 1935 to 1938 parts of the *autobahn* network had been finished, bus routes on the new segments were opened immediately after the opening of the new *autobahn* segments. These bus routes were run by the *Reichsbahn*. As the operating company of the *autobahn*, the *Reichsautobahngesellschaft*, has been established as a daughter of the *Reichsbahngesellschaft*, the latter company took the responsibility for generating the urgently needed traffic on the *autobahn* which exhibited only minor traffic so far. And as an additional goal the bus routes supported tourist excursions. As a third goal, the buses provided tours on the *autobahn* for those people, who could not afford an own car. So, the buses served as an instrument of propaganda to show the people the giant roads. The right to run bus routes on the *autobahn* was given exclusively to the *Reichsbahn*[701] with the consequence that private entrepreneurs were excluded from a new attractive segment of the growing service economy. This consideration fits to the Nazi economic policy to restrict the economic activities of the small business sector, as I have demonstrated on the example of truck entrepreneurs.[702] This policy to crowd private enterprises out of the bus industry differed totally from the traffic policy in the United States.[703]

The first bus route on the *autobahn* Frankfurt a.M.–Darmstadt started operation from main station Frankfurt a.M. to main station Darmstadt on 20th May 1935, one day after the opening of this route.[704] To stress the modernity of the new transport mode, special buses with streamlined design were procured. The bus ride took 38 minutes and lasted 17 minutes longer

---

701 See the lecture of Rudolf Hoffmann, an official in Todt's administration: Die Wirtschaftlichkeit von Autobahnen, in: Deutsche Bergwerkszeitung on 23 November 1938, Bayerisches Wirtschaftsarchiv Munich, K9/1491. Already in 1933 the head of the construction project company for the autobahn route Munich–Salzburg, the delegated railway official Doll, raised in a letter on 11 December the question, how the German Railway Company could furnish the bus traffic which could be expected on the new autobahn, in: Bavarian State Archive Munich, Autobahndirektion Süd, vol.248. See also Stern.

702 Herbst, Ludolf: Der Totale Krieg und die Ordnung der Wirtschaft, Stuttgart 1985, p.153–160. The discrimination of travel agencies is explored by Wolfgang König: Nazi Visions of Mass Tourism, op. cit. The discrimination of driving schools is shown by Dorothee Hochstetter: Motorisierung und Volksgemeinschaft, München 2005. Vahrenkamp 2008.

703 Magaret Walsh: Making Connections. The Long Distance Bus Industry in the United States, Aldershot 2000.

704 The article Der Kraftwagenschnellverkehr der Reichsbahn, in: Deutsche Bergwerkszeitung on 24 May 1935, Bavarian Economic Archive Munich, K9/1491, stresses the tourist importance of this bus link.

than a train passage. In addition, the price of a ticket was above a train ticket in the 3rd class. So, the bus route did not offer reasonable conditions and did not attract many passengers.[705]

Other examples of bus routes on the *autobahn* can be found in the archives.[706] The first example is the bus route Kassel–Frankfurt a.M. from main station to main station with two bus pairs each day. The duration of the bus ride amounted to 2 hours and 50 minutes and was comparable with a corresponding journey by train with medium speed (Eilzug – local train). The bus route opened on 4th of December 1938, one day after the opening of the entire autobahn route Kassel–Frankfurt a.M.

The ticket price for the bus was also comparable to the price of the express train ticket. In a report about a bus ride in January 1939 by the NS party press published in the *Kurhessische Landeszeitung* newspaper, the reporter especially emphasized the touristic highlights of this route, such as the Neuenstein Castle and how the route was adapted to the landscape.

Image 71: A Henschel bus in front of the Neuenstein Castle

705 Walter Linden: Reichsbahn-Autobus-Verkehr auf Reichsautobahnen, in: Deutsche Wirtschaftszeitung on 23 May 1935, p. 480–483, and R. Roesner: Der Kraftwagen-Linienverkehr der Reichsbahn auf den Reichsautobahnen, in: Deutsche Wirtschaftszeitung on 23 April 1936 p. 386–389, both Bayerisches Wirtschaftsarchiv K9/1491.

706 There are documents for other bus routes in 1939, e.g. Munich–Augsburg, Munich–Nuremberg–Berlin, SM, vol. Verkehr 188.

Image 71 shows the Henschel bus in front of the Neuenstein Castle.[707] It was emphasized that the bus ride brings travelers closer to the landscape than a train ride does. The following Image 72 is taken from this report and shows the Mercedes bus driving on the empty expressway in January 1939.[708]

Image 72: A Mercedes bus on the low-traffic stretch Kassel–Frankfurt

When the *autobahn* route Kassel–Göttingen was opened on June 20, 1937, a bus connection from Kassel to Göttingen was announced but only began operation with the winter schedule beginning on October 2, 1937. According to a report in the *Kasseler Post* newspaper, streamlined express buses were to be used just as on the expressway line Frankfurt–Darmstadt.[709]

---

707 Rolf Mengel: Die schnellen Riesen der Reichsautobahn, in: Kurhessische Landeszeitung 14 January 1939, p. 3.

708 l.c.

709 Kasseler Post 10 June 1937 and Die Autobahn, issue 19, 1934.

Image 73: Henschel advertisement for streamliners[710]

Image 73 shows several streamlined buses of the Henschel company – with a chassis from Credé (model 38 S3N) – under the Werra Valley expressway bridge. The streamlined shape fits into the debate on rationalization, power and speed of the 1920s.[711] Two of these Henschel buses traveled daily between the main train stations of Kassel und Göttingen. There were two departure times at Kassel: 9:00 am and 8:50 pm, and the buses also stopped at the Old Market (Altmarkt). The duration of the ride was one hour, making it competitive to the train connection. A passenger train on the stretch Kassel–Göttingen required one hour and 37 minutes.[712] Image 74 shows a Henschel bus on the Werra Valley Bridge.[713] Following a suggestion made by the *Kasseler Post*, discounted tickets were sold on Sundays for the stretch Kassel–Göttingen in order to give excursionists the chance to get to know the new stretch.[714]

---

710 Stern, p. 42.

711 Peter Borscheid: Die Tempomacher: Die Rationalisierungsbewegung und die Beschleunigung des Lebens in den Weimarer Jahren, in: Zeitschrift für Unternehmensgeschichte, vol. 41, 1996, pp. 125–138.

712 Reichsbahn timetable 1939.

713 Henschel Museum Kassel.

714 Kasseler Post on 26 July 1937.

Image 74: A Henschel bus on the Werra Valley expressway bridge

The system of express buses on the expressways of the Reich was successively expanded. Traveling daily from Frankfurt a.M. in 1937, there were three pairs of buses for the route Homburg–Bad Nauheim–Friedberg, 14 pairs of buses to Darmstadt, five for the route Mannheim–Karlsruhe–Pforzheim–Stuttgart, three pairs of buses for the route Darmstadt–Heidelberg and seven pairs of buses for the route Mannheim–Heidelberg.[715]

Immediately following the inauguration of the first Munich–Holzkirchen leg on June 29, 1935, bus lines connecting Munich to Tegernsee and Munich to Schliersee were opened at the Holzkirchen junction.[716] The three daily connections that were offered from the Starnberg train station in Munich operated according to the schedule shown on Table 13.[717] The schedule shows that a bus trip from Munich to Tegernsee took 1 hour and 45 minutes – a traveling time

715 Schedule valid beginning 15 February 1937, Magistratsakten MA 6289, ISG.

716 Münchener Zeitung on 28 June 1935, Stadtarchiv, Munich, München–Salzburg 1935. In this archive there are additional documents on bus routes in 1939, for example, Munich–Augsburg and Munich–Nuremberg–Berlin, see volume Verkehr 188.

717 Münchener Zeitung 28 June 1935,City Archive, Munich, München–Salzburg 1935.

that equaled the traveling time by train to this destination. A train schedule from 1939 shows that a train trip from Munich to Scharflach took 1 hour and 2 minutes while the local train traveled to Tegernsee in 24 minutes.[718]

	First Bus	Second Bus	Third Bus
Bus line to Tegernsee:			
Munich departure	10:00	17:50	23:15
Tegernsee arrival	11:15	19:05	00:30
Tegernsee departure	6:45	15:35	21:00
Munich arrival	8:00	16:50	22:15
Bus line to Fischhausen:			
Munich departure	9:50	18:00	23:10
Schliersee arrival	11:18	19:28	00:38
Fischhausen arrival	11:27	19:37	00:47
Fischhausen departure	6:30	15:23	20:33
Schliersee departure	6:39	15:32	20:42
Munich arrival	8:07	17:00	22:10

Table 13: Schedule of Bus Lines München–Tegernsee and Schliersee

Traffic censuses at the expressway near Ramersdorf (the expressway access point from the Munich suburb) indicate that the number of buses per month in 1937 and 1938 fluctuated between 50 and 150. Most of these buses constituted unscheduled excursion traffic rather than regular bus line traffic. The following diagram shows the development of bus traffic on the expressway.[719]

[718] Siegfried Bufe et al.: Die Tegernsee Bahn, Egglham 2001, p.24.

[719] Rudolf Hoffmann: Die Wirtschaftlichkeit von Autobahnen, op. cit.

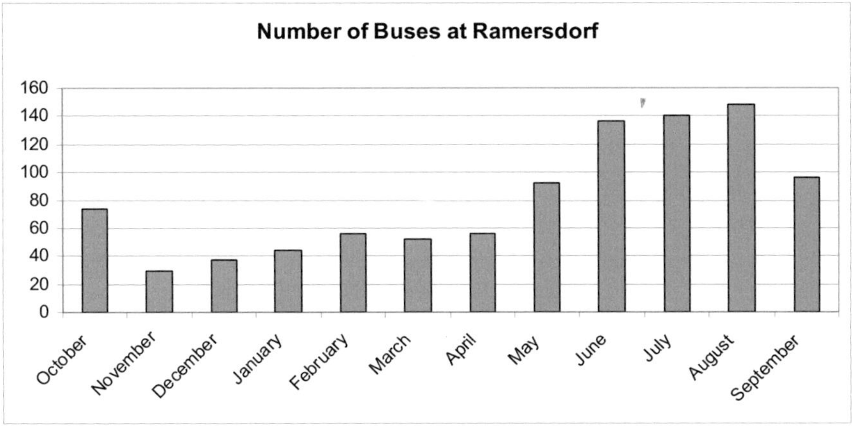

Diagramm: Number of Buses at Ramersdorf from October 1937 to September 1938

The *Deutsche Reichsbahn*, however, was soon forced to limit bus operations on the expressways in order to save economic resources, which were to primarily flow into the four-year rearmament plan.[720] This brought about considerable tensions between the interests of a NS consumer society and the military goal of rearmament.

[720] Letter from the Reichsbahn, Direction Frankfurt am Main to the mayor of Frankfurt on 12 February 1937, Magistratsakten MA 6289, ISG.

26 Bibliography

Contemporary Sources

ADAC (ed.) Reichs-Verkehrsführer und Reichs-Tourenbuch, Dortmund 1930.

ADAC-Motorwelt (1930): Ausbau der Fernstrassen in Deutschland, ADAC-Motorwelt, issue 38, vol. 25, pp. 27–29.

Almeida, Ludwig, Graf: (1929): Kraftfahrer und Nur-Autostraße, Hafraba-Mitteilungsblatt, issue 11, 3, pp. 4–5.

Auberlen, R.: Die Verbindung des Landstrassennetzes mit der Reichsautobahn Karlsruhe–Frankfurt–Hersfeld, in: Die Strasse, issue 11, 1937, pp. 302–307.

Autofernstrassen-Conference Leipzig (1927): Autofernstraße Berlin–Leipzig–München–Rom, Neues Rathaus, Verlag Rats-Verkehrsamt, Leipzig.

Becker, Kurt: Göttingen–Hann. Münden–Kassel, in: Hafraba-Mitteilungsblatt Nr. 2, 1931, pp. 1–4.

Bericht (1926): Bericht über die Besprechung des Projekts Automobilstrasse Hamburg–Mailand, Druckschrift des Wirtschaftsamts Frankfurt, Akte Städtisches Verkehrsamt, 10/3, Stadtarchiv Kassel.

Bretz, Hans: Das Rheinland baut Strassen, in: ADAC Motorwelt Nr. 33, 1932, p. 27s.

Cuno, Wilhelm (1929): Die Hafraba und der Fremdverkehr von Übersee, Hafraba-Mitteilungsblatt, issue 4, pp. 2–3.

Derboven, Fritz: Der Güterverkehr des Düsseldorfer Wirtschaftsgebietes, Diss. Universität Köln 1929.

Dinklage, Albert (1934): Autobahnen verlangen andere Motoren, Die Autobahn, issue 14, 3, p. 638.

Dirks, Walter: Das Dreieck auf den Autobahnen – Impressionen einer Fahrt Frankfurt–Berlin–München–Frankfurt, in: Walter Dirks: Feuilletons im Nationalsozialismus, Gesammelte Schriften, volume 3, edited by Fritz Boll, Ulrich Bröckling and Karl Prümm, Zürich 1990, pp. 136–152.

Doll, Friedrich: Vom Bau der Reichsautobahn München–Landesgrenze, in: Die Strasse, issue 5, 1935, pp. 153–154.

Ewald, Kurt: 125 Jahre Henschel, Kassel 1935.

Frey, Friedrich (1928): Das Neue Strassenverkehrsrecht im Reich und in den Ländern, Mainz.

Frey, Friedrich (1929): Strassenwesen und Strassenverkehrswesen im Rhein-Main-Gebiet, Habraba-Mitteilungsblatt, issue 11, 3, p. 6.

Generalinspekteur für das deutsche Strassenwesen (ed.): Der Kraftverkehr auf Reichsautobahnen, Reichs- und Landstrassen im Dritten Reich, Berlin 1939.

Gesellschaft zur Vorbereitung der Reichautobahnen (ed.): Die Planungsarbeiten für die Reichsautobahnen – Zweieinhalb Jahre GEZUVOR, Berlin 1937.

Golder: Weiterausbau der Hafraba-Idee, in: Hafraba-Mitteilungsblatt, issue 2, 1930, pp. 1–3.

Grossjohan (1929): Die Kraftwagenstrasse Bonn–Köln–Düsseldorf, in: Strassenbau und Strassenunterhaltung, Beilage der Zeitschrift Verkehrstechnik, 10, p. 416.

Grossjohann: Der Beginn einer neuen Epoche in Deutschland – Zur Eröffnung der Autobahn Köln–Bonn am 6. August 1932, in: Die Autobahn, issue 8, 1932, pp. 1–5.

Grüb, Anton: Reichsautobahn-Raststätte am Chiemsee, in: Die Strasse, 1938, issue 7, pp. 217–218.

Guderian, Heinz: Mit der Panzerwaffe auf den Straßen des Sieges, in: Die Straße, issue 23/24, 1940, pp. 504–506.

Heller, Kurt (1928): Das Kraftwagenstrassennetz Deutschlands, in: Strassenbau und Strassenunterhaltung, supplement to the journal Verkehrstechnik, in: Verkehrstechnik, 9, pp. 669–671.

Hoffmann, Karl (1931): Die Ergebnisse der 2. Deutschen Verkehrszählung von 1928–29, Verkehrstechnik, 12, pp. 189–191.

Hoffmann, Rudolf: Der Kraftverkehr auf deutschen Straßen Ostern 1939, in: Die Strasse, issue 9, 1939, pp. 283–289.

Hoffmann, Rudolf: Der Kraftverkehr auf deutschen Strassen zu Pfingsten 1938, in: Die Strasse, issue 12, 1938, pp. 388–392.

Hoffmann, Rudolf: Der Kraftverkehr auf deutschen Strassen zu Pfingsten 1939, in: Die Strasse, issue 11, 1939, pp. 377–380.

Hoffmann, Rudolf: Die Entwicklung des Kraftverkehrs auf Reichsautobahnen und Reichsstrassen, in: Forschungsgesellschaft für das Strassenwesen (ed.): Strassenbau-Jahrbuch 1937–1938, Berlin, Verlag Volk und Reich, 1938, pp. 29–42.

Horn, J.: Autobahn und Industrie, in: Hafraba-Mitteilungsblatt, issue 7, 1931, p. 1s.

Institut Für Konjunkturforschung (1930): Lastkraftwagen in der Volkswirtschaft, weekly report on 7 May 1930, pp. 23–24.

Kaftan: Hafraba – Ein Rückblick, in: Die Strasse, issue 9, 1935, p. 321s.

Kamm, W. and H. Goll: Einfluß der Reichsautobahn auf die Gestaltung der Kraftfahrzeuge, in: Die Strasse, issue 3, 1936, p. 67s.

Kappler, Carl: Landstrassen in Deutschland, in: Hafraba-Mitteilungsblatt, issue 9, 1930, pp. 1–2.

Kirchner, Ingeborg: Die Industrielle Entwicklung der Stadt Kassel 1866–1914, Diss. Cologne 1921

Krahl, Otto: Strassen entstehen ...Henschel hilft bauen, in Henschel-Stern, issue 2, 1936.

Kruse, H. : Die Notlage der Kraftverkehrswirtschaft, in: Auto und Wirtschaft, Nr. 1, Berlin 1932, p. 1.

List, Friedrich: Über ein sächsisches Eisenbahnsystem als Grundlage eines allgemeinen deutschen Eisenbahnsystems, und insbesondere über die Anlegung einer Eisenbahn von Leipzig nach Dresden, Leipzig 1833.

Lüer, Carl: Neue Möglichkeiten der Lebensmittelversorgung durch den Kraftverkehr mit besonderer Berücksichtigung der Reichsautobahn, in: Die Strasse, 1935, issue 9, pp. 334–339.

Lutz, Oswald: 3000 Kilometer Reichsautobahn für den gewerblichen Kraftverkehr, in: Die Strasse, 1938, issue 24, p. 772.

Mellerovicz, Konrad: Autobahnen und Kraftverkehrswirtschaft, in: Die Autobahn, issue 9, 1934, p. 370.

Mellerowicz, Konrad: Der Stand des gewerblichen Güterfernverkehrs, in: Die Strasse, issue 5, 1935, pp.143–147.

Meyer, Julius and Heinrich Wilkin: Die Landstraße als Heimat des Lastzugbesitzers, in: Die Straße, 1935, pp.160–163.

Neumann (1931): Strassenverkehr und Strassenwesen in den USA, in: Habraba-Mitteilungsblatt, issue 1, vol. 5, p. 4.

Nyfeller, Marcel (1930): Das Projekt einer Autostrasse Bern–Thun, in: Hafraba-Mitteilungsblatt, issue 12, vol. 4 p. 6.

Örley, Leopold (1935): Die Großglockner-Hochalpenstraße, in: Die Strasse, issue 10, vol. 2, pp. 372–374.

Otzen, Robert (1927): Die Autostrasse Hansestädte–Frankfurt–Basel, Ziele und Zweck des Vereins 'Hafraba', Hannover.

Otzen, Robert (1931): Zur Systematik des deutschen Landstrassenbaus, Bautechnik, issue 26, Akte 1/H/3500, 25, Stadtarchiv Karlsruhe.

Otzen, Robert: Drei Jahre Hafraba, in: Hafraba-Mitteilungsblatt, issue 11, 1929, p. 3

Pflug, Fritz (1928): Der Kraftfahrzeugverkehr, in: Bernd Harms (ed.), Strukturwandlungen der Deutschen Volkswirtschaft, Berlin, vol. 2, pp. 248–265.

Pohl, Heinrich: Die Reichsbahn im Raum Köln. Diss. Universität Köln 1933.

Puricelli, P.: Entwurf für ein europäisches Autostrassennetz, in: Die Strasse, Heft 2, 1934, p. 42s.

Quatram, Josef: Verkehrs- und wirtschaftspolitische Vorbedingungen für den Bau einer Schnellbahn im rheinisch-westfälischen Industriegebiet mit besonderer Berücksichtigung des Schnellbahnprojekts Köln–Dortmund, Diss. Universität Köln 1926.

Rappaport (1926): Das Netz der Kraftwagenstrassen Deutschlands, Verkehrstechnik, 7, pp. 285–288.

Reichsverkehrsministerium (Ed.): Hundert Jahre deutsche Eisenbahnen, second edition, 1938.

Reichszentrale Für Deutsche Verkehrswerbung (1926): Deutscher Verkehrsdienst, No. 289, Stadtarchiv München, Verkehr 84.

Reifferscheidt, Adolph: Der Nahverkehr im Kölner Wirtschaftsbezirk, Diss. Universität Köln 1926.

Reyer, Ernst: Reichsautobahn Kassel–Göttingen, in: Henschel-Stern, issue 4, 1936, pp.72–76

Rheinische Provinzialverwaltung (ed.): Die Kraftwagenstrasse Köln–Bonn, Düsseldorf, 1932, Stadtmuseum Bonn.

Rouselle, U. (1931): Zur Belebung der Bauwirtschaft unter Berücksichtigung der Hafraba, Habraba-Mitteilungsblatt, issue 7, vol. 5, p.5.

Scherl, August: Ein neues Schnellbahnsystem – Vorschläge zur Verbesserung des Personenverkehrs, Berlin 1909.

Schmidt, Armin: Die Verkehrsentwicklung Kassel während der letzten 100 Jahre, Diss. Göttingen 1927.

Schmidt-Lamberg, Herbert: Hebung des Fremdenverkehrs durch Autobahnen, in: Die Autobahn, issue 6, 1933, p. 2.

Schmitt: Die Milchversorgung der Großstädte, in: Der Güterumschlag, Sonderausgabe der Zeitschrift des Vereins Deutscher Ingenieure, Berlin, 1926, pp. 235–237.

Scholz, W.: Die Entwicklung des Kraftfahrzeugbestandes im gewerblichen Güterfernverkehr, in: Die Strasse, issue 4, 1938, pp. 106–108.

Schönleben, Eduard: Das Netz der Reichsautobahnen, in: Die Strasse, issue 19, 1936, pp. 614–616.

Schönleben, Eduard: Fritz Todt, Oldenburg, 1943.

Sommerer, Wilhelm: Ausbau der Reichsstraße zu den Kampfstätten der IV. Olympischen Winterspiele (Reichsstraße Nr. 2, München–Garmisch) und Verkehrsleistung während der Spiele, in: Die Strasse, issue 5, 1936, pp. 156–157.

Speck, Arthur (1932): Der Fernverkehr auf deutschen Landstrassen, Verkehrstechnik, vol. 13, pp. 239–242.

Studiengesellschaft für die Finanzierung des Deutschen Strassenbaus (ed.) (1930): Die Finanzierung des deutschen Strassenbaus, Berlin.

Tagungsbericht Der 1. Verwaltungsrat-Sitzung (1927): Schrift Nr. 2 on 10 February 1927, Hellerdruck Frankfurt.

Thalenhorst (1930): Städte an den Hafraba-Strassen: Bremen, Habraba-Mitteilungsblatt, issue 4, vol. 4, pp. 1–3.

Todt, Fritz (1936): Reichsautobahnen in der Wirtschaft, Frankfurter Volksblatt 27 February 1936, Archiv Hessischer Rundfunk.

Todt, Fritz: Bau und Verkehr auf den Strassen Adolf Hitlers, in: Die Strasse, issue 18, 1937, pp. 518–520.

Todt, Fritz: Der Generalinspekteur für das deutsche Strassenwesen berichtet auf dem Parteitag Großdeutschlands über die Strassen des Führers als Schrittmacher der deutschen Bauwirtschaft, in: Die Strasse, Heft 19, 1938, pp. 600–603.

Uhlfelder, H. (1930): Die oberitalienischen Autobahnen, Hafraba-Mitteilungsblatt, vol. 4, p. 11.

Uhlfelder. H. (1931): Richtlinien für die technische Anlage von Autobahnen, Teil C: Abzweigungen und Kreuzungen, Hafraba-Mitteilungsblatt, issue 11, vol. 5, pp. 2–7.

Uhlfelder: Richtlinien für den technische Anlage von Autobahnen, Teil A: Strecken im Flachland, in: Hafraba-Mitteilungsblatt 4/1931, pp. 4–6.

Uhlfelder: Richtlinien für die technische Anlage von Autobahnen, Teil B: Stationen, in: Hafraba-Mitteilungsblatt, Heft 10, 1931, pp. 6–8.

van Elsen, Paul: Die deutsche Landstrasse, Diss. Cologne, Düren 1929.

Vilbig: Ausbau der Straßen – Nur-Autostraßen? Drei Musterbeispiele aus Bayern, in: Verkehrstechnik, issue 21, pp. 327–330.

Vollbehr, Ernst: Arbeitsschlacht – Fünf Jahre Malfahrten auf den Bauplätzen der Strassen Adolf Hitlers, Berlin, 2nd edition, 1938.

Woerner, Anton: Die Reichsautobahn München–Salzburg, in: Die Autobahn, issue 9, 1933, pp.10–12.

Literature

Adam, Werner: Entstehung und Bau der Reichsautobahn, in: Festschrift zur 600 Jahr Feier Guxhagen, Guxhagen 1952, pp. 65–71.

Adenauer, Konrad: Konrad Adenauer als Präsident des Preußischen Staatsrates, in: Hugo Stehkämper (ed.): Konrad Adenauer. Oberbürgermeister von Köln, Köln 1976, pp. 335–404.

Attali, Jean: The Roman System, or the Generic in All Times and Tenses, in: Rem Koolhaas (ed.): Mutations, Barcelona 2000

Auberlen, R.: Die Verbindung des Landstrassennetzes mit der Reichsautobahn Karlsruhe–Frankfurt–Hersfeld, in: Die Strasse, issue 11, 1937, pp. 302–307.

Baldwin, Neil: Henry Ford and the Jews, New York 2001.

Baranowski, Shelley: Strength through Joy: Consumerism and Mass Tourism in the Third Reich, New York 2004.

Baranowski, Shelley: Tourism Industry and Modernity: Comments, in: Laurent Tissot (ed.): Development of a Tourist Industry, Neuchatel 2003.

Barth, Klaus-Peter and Michael Antenbrink: 50 Jahre Frankfurter Kreuz – Von der Nur-Autostrasse zum europäischen Verkehrsknotenpunkt, in: Strassen-Verkehrstechnik, issue 6, 2006, pp. 341–348.

Baumann, Carl-Friedrich: 175 Jahre Henschel, Moers 1985.

Becker, Kurt, Frank: Autobahnen, Auto-Mobilität. Die USA; Italien und Deutschland im Vergleich, in: Politische Kulturgeschichte der Zwischenkriegszeit 1918–1939, Göttingen 2005 (Geschichte und Gesellschaft Sonderheft 21), pp. 23–59.

Bönig, Jürgen: Die Einführung von Fließbandarbeit in Deutschland bis 1933 – Zur Geschichte einer Sozialinnovation, Frankfurt 1993.

Bonino, Michele and Massimo Moraglio: Inventing Movement – History and Images of the A6 Motorway, Turin 2006.

Borscheid, Peter: Die Tempomacher: Die Rationalisierungsbewegung und die Beschleunigung des Lebens in den Weimarer Jahren, in: Zeitschrift für Unternehmensgeschichte, vol. 41, 1996, pp. 125–138.

Bortolotti, Lando (1992): Origine e Primordi della Rete Autostradale in Italia 1922–1933, Storia Urbana, vol. 59, pp. 35–69.

Bortolotti, Lando (1997): Fra Politica, Propaganda e Affari: L'Autostrada Roma–Berlino 1927–1942, Storia Urbana, vol. 81, pp. 47–79.

Bortolotti, Lando and Guiseppe De Luca: Fascismo e autostrada, Milano 1994.

Bufe, Siegfried et al.: Die Tegernsee Bahn, Egglham 2001.

Bundesminister für Verkehr (ed.): Die Verkehrspolitik in der Bundesrepublik Deutschland 1949–1953, Dortmund 1953.

Bundesminister für Verkehr (ed.): HAFRABA – Bundesautobahnen Hansestädte–Frankfurt–Basel, Bau Verlag, Wiesbaden und Berlin 1962.

Churella, Albert: From Steam to Diesel. Mangerial Customs and Organizational Capabilities in the Twentieth-Century Armerican Locomotive Industry, Priceton 1998.

Clark, Gordon (ed.): Oxford Handbook of Economic Geography, Oxford 2003.

Cohausz, Otto: Der französisch-deutsche Eisenbahnkrieg im Ruhrgebiet 1923, in: Jahrbuch für Eisenbahngeschichte, Bd. 6, 1973, pp. 5–25.

Corni, C.: Hitler and the Paesants, New York 1990.

Darchinger, Josef: Wirtschaftswunder – Deutschland nach dem Krieg 1952–1967, Köln 2008.

Denner, Karl: Planung und Entwurf der Strecke Kassel–Frankfurt–Mannheim, pp. 97–100, in: Bundesminister für Verkehr (ed.): Hafraba – Bundesautobahnen Hansestädte–Frankfurt–Basel, Bau Verlag, Wiesbaden und Berlin 1962.

Domarus, M: Hitler – Reden und Proklamationen, volume I, Wiesbaden 1973.

Edelmann, Heidrun (1989): Vom Luxusgut zum Gebrauchsgegenstand: die Geschichte der Verbreitung von Personenkraftwagen in Deutschland, Frankfurt.

Edelmann, Heidrun: Der Traum vom "Volkswagen", in: Hans-Liudger Dienel and Helmuth Trischler (eds.): Geschichte der Zukunft des Verkehrs, Frankfurt 1997, pp. 280–289.

Eichner-Ramm, Britta: 70 Jahre Autobahn Göttingen–Kassel, Zeitzeugen berichten, Göttingen, 2007.

Eisenbach, Ulrich and Gerd Hardach (eds.): Reisebilder aus Hessen, Fremdenverkehr, Tourismus und Kur seit dem 18. Jahrhundert, Darmstadt 2001.

Erdmann, Karl Dietrich (ed.): Akten der Reichskanzlei Weimarer Republik, Kabinette Brüning I a. II, volume 1, 30.März 1930 bis 10.Oktober 1931, edited by Tilmann Koops, Boppard am Rhein, 1990.

Fahle, G.: Nazis und Bauern, Köln 1986.

Flik, Reiner (2001): Von Ford lernen? Automobilbau und Motorisierung in Deutschland bis 1933, Köln.

Flink, James: America Adopts the Automobile, 1895–1910, Cambridge, Mass., 1970.

Flyvbjerg, Bent; Bruzelius, Nils; Rothengatter, Werner: Megaprojects and risk: an anatomy of ambition, Cambridge: Cambridge Univ. Press, 2003.

Foster, Mark: From streetcar to Superhighways: American City Planners and Urban Transit 1900–1940, Philadelphia 1981.

Frielingsdorf, Volker: Konrad Adenauers Wirtschaftspolitik als Kölner Oberbürgermeister (1917–1933), Basel, 2002.

Fröhlich, Elke (ed.): Die Tagebücher von Josef Goebbels, volume 2, Part 1, München 1987.

Gabriel, Roland: Deutsche Autobahnvorläufer – AVUS, Hafraba, München–Leipzig–Berlin, Leipzig–Halle, Düsseldorf–Köln–Bonn, in: Wolfgang Wirth (ed.): Die Autobahn – Von der Idee zur Wirklichkeit, Köln 2005, pp. 24–62.

Gottwaldt, Alfred: Julius Dorpmüller, die Reichsbahn und die Autobahn, Berlin 1995.

Gottwaldt, Alfred: Neuartige Dampflokomotiven von Henschel in Kassel. Das Lebenswerk des Ingenieurs Richard Roosen. Ein Beitrag zur Industriegeschichte Nordhessens 1925–1957, in: Zeitschrift des Vereins für hessische Geschichte und Landeskunde, 106 (2001), pp. 299–308.

Grundmann, F.: Agrapolitik im Dritten Reich, Hamburg 1979.

Haas, Robert et al.: Weltbühne und Naturkulisse – Zwei Jahrhunderte Salzburg-Tourismus, Salzburg 1994.

Haerendel, Ulrike and Gabriele Krüger: „Groß-München“: Eingemeindung, Verkehr, kommunales Bauen, in: München-Hauptstadt der Bewegung, edited by Stadtmuseum Munich, 2002, pp. 287–293.

Hafen, Paul (1956): Das Schrifttum über die deutschen Autobahnen, Bonn, pp. 30–32.

Hahn, Jürgen: 50 Jahre Autobahnen in Deutschland, in: Strasse und Autobahn, issue 1, 1983, pp. 1–4.

Hamilton, Shane: Trucking Country, Princeton UP 2008.

Hardach, Gerd and Eisenbach, Ulrich (eds.): Reisebilder aus Hessen, Fremdenverkehr, Tourismus und Kur seit dem 18. Jahrhundert, Darmstadt 2001.

Heckmann-Strohkark, Ingrid: Der Traum von einer europäischen Gemeinschaft. Die internationalen Autobahnkongresse 1931 und 1932, in: Martin Heller and Andreas Volk (eds.): Die Schweizer Autobahn, Zürich, 1999.

Heller, Martin and Andreas Volk (eds.): Die Schweizer Autobahn, Zürich, 1999.

Hellmuth, Th. and E. Hiebl: Tourismusindustrie – Organisation und quantitative Entwicklung des Salzburger Fremdenverkehrs 1860–1938, in: Haas, Robert et al.: Weltbühne und Naturkulisse – Zwei Jahrhunderte Salzburg-Tourismus, Salzburg 1994.

Henning, H.: Kraftfahrzeugindustrie und Autobahnbau in der Wirtschaftspolitik der Nationalsozialisten 1933 bis 1936, in: Vierteljahresschrift für Sozial- und Wirtschaftsgeschichte, Bd. 65, 1978, pp. 217–242,

Herbert, Ulrich: Fremdarbeiter – Politik und Praxis des "Ausländer-Einsatzes" in der Kriegswirtschaft des Dritten Reiches, Bonn 1999

Herbst, Ludolf: Der Totale Krieg und die Ordnung der Wirtschaft, Stuttgart 1985.

Herf, Jeffrey: Reactionary Modernism. Technology, Culture, and Politics in Weimar and the Third Reich, Cambridge, 1984.

Herz, Rudolf and Reinhard Matz: Zwei Entwürfe zum Holocaust-Denkmal, Nürnberg 2001.

Historische Kommission bei der Bayerischen Akademie der Wissenschaften (eds.): Akten der Reichskanzlei Online, Das Kabinett Marx III/IV.

Hochstetter, Dorothee: Motorisierung und Volksgemeinschaft, München 2005.

Hohlmann, Susanne: Pfaffenwald – Sterbe- und Geburtenlager 1942–1945, Nationalsozialismus in Nordhessen vol. 2, Kassel 1983.

Holzer, Gabriele: Verfreundete Nachbarn Österreich – Deutschland, Wien 1995.

Jacobsen, Hans-Adolf: Der Fall Gelb, Wiesbaden 1957.

Joerges, Bernd: Die Brücken des Robert Moses: Stille Post in der Stadt- und Techniksoziologie, in: Leviathan, 1999, issue 1.

Johannes, Ralph and Gerhard Wölki: Die Autobahn und ihre Rastanlagen – Geschichte und Architektur, Petersberg 2005.

Kaftan, K.: Der Kampf um die Autobahn, Berlin 1955.

Kästner, Erich: Notabene 45, Zurich 1961.

Keitz, Christine: Reisen als Leitbild – Die Entstehung des modernen Massentourismus in Deutschland, Munich 1997.

Kershaw, Ian: Der NS-Staat, Hamburg 1988.

König, Wolfgang: Nazi Visions of Mass Tourism, in: Laurent Tissot (Ed.): Development of a Tourist Industry, Neuchatel 2003, pp. 261–268.

König, Wolfgang: Volkswagen, Volksempfänger, Volksgemeinschaft – Vom Scheitern der nationalsozialistischen Konsumgesellschaft, Paderborn 2004.

Kopper, Christopher: Modernität oder Scheinmodernität nationalsozialistischer Herrschaft – Das Beispiel der Verkehrspolitik, in: Von der Aufgabe der Freiheit, Festschrift für Hans Mommsen, edited by Christian Jansen, Lutz Niethammer, and Bernd Weisbrod, pp. 399–411, Berlin: Akademie Verlag, 1995.

Kopper, Christopher: Handel und Verkehr im 20. Jahrhundert, München, 2002.

Kornrumpf, Martin: HAFRABA e.V. – Deutsche Autobahn-Planung 1926–1934, Bonn 1990.

Kreuzer, Bernd: Tempo 130. Kultur- und Planungsgeschichte der Autobahnen in Oberösterreich, Linz 2005.

Krüger, P.: Zu Hitlers ‚nationalsozialistischen Wirtschaftskenntnissen', in: Geschichte und Gesellschaft, vol. 6, 1980, pp. 263–282.

Kuhlmann, Walter: GEZUVOR – Was geht vor? Betonstrecke oder Ackerland, in: Heimatjahrbuch Lahn-Dill Kreis, 2008, pp. 144–152, Bad Ems, 2007.

Lang, K. and R. Stommer: Deutsche Künstler – an die Front des Strassenbaues!, in: Stommer and Philipp: Reichsautobahnen – Pyramiden des Dritten Reiches, Marburg 1982.

Lärmer, Karl: Autobahnbau in Deutschland 1933 bis 1945, Berlin 1975.

Ludwig, Karl-Heinz: Politische Lösungen für technische Innovationen 1933–1945 – Eine antitechnische Mobilisierung, Ausformung und Instrumentalisierung der Technik, in: Technikgeschichte, Bd. 62, 1995, pp. 333–344.

Ludwig, Karl-Heinz: Technik und Ingenieure in Dritten Reich, Königstein 1974.

Lüken-Isberner, Folckert (1990): Fritz Stück und die Stadtplanungsdiskussion im Kassel der Weimarer Republik, Hessisches Jahrbuch für Landesgeschichte, 40, pp. 219–240.

Marcon, H.: Arbeitsbeschaffungspolitik der Regierungen, Papen and Schleicher, Frankfurt 1974.

Mason, Timothy W.: Arbeiterklasse und Volksgemeinschaft, Gütersloh 1975.

Minuth, Karl-Heinz (ed.): Records of the Reich Chancery, Das Kabinett von Papen, volume 1, Boppard on the Rhine 1989.

Mom, Gijs (2005): Roads as Rails – European Highway-Network Building and the Desire for Long-Range Motorized Mobility, Technology & Culture, vol. 46, pp. 745–772.

Mommsen, Hans and Mandfred Grieger: Das Volkswagenwerk und seine Arbeiter im Dritten Reich, Düsseldorf 1996.

Moraglio, Massimo (2002): Una storia delle autostrada italiane: il periodo fra le guerre mondiali, in: Storia urbana, vol. 100, pp. 11–26.

Nietfeld, Anette: Reichsautobahn und Landschaftspflege, Werkstattberichte des Instituts für Management und Umweltplanung, edited by Arnim Bechmann, Nr. 14, 1985, TU Berlin.

Overy, Richard: Interrogations – The Nazi Elite in Allied Hands, London 2001.

Patton, Phil: Open Road – A Celebration of the American Highway, New York 1986.

Picon, Antoine: Towards a history of Infrastructure, in: Michele Bonino und Massimo Moraglio: Inventing Movement – History and Images of the A6 Motorway, Turin 2006, pp. 13–17.

Piwecki, Frank: Frühe Konzepte des individuellen Massenverkehrs in Deutschland, dargestellt am Beispiel des Vereins "HAFRABA", Magisterarbeit im Hauptfach Geschichte, Universität Hannover 1996.

Pohl, Hans, Stephanie Habeth and Beate Brüninghaus (1986): Die Daimler-Benz AG in den Jahren 1933 bis 1945. Eine Dokumentation, Zeitschrift für Unternehmensgeschichte, Beiheft 47, Stuttgart.

Pohl, Manfred: Die Strabag 1923 bis 1998, München 1998.

Pusch, Wolfgang: Die „Entdeckung“ des Würm- oder Starnberger Sees und die Folgen, in: Landpartie. Museen rund um München, published by Landpartie, Munich 2002, pp. 180–195.

Radde, B: The Merrit Parkway, Yale University Press 1996.

Rauscher, Walter (2001): Hitler und Mussolini, Graz.

Reichsverkehrsministerium (Ed.): Hundert Jahre deutsche Eisenbahnen, second edition, Berlin 1938.

Repgen, Konrad (ed.) (1983): Akten der Reichskanzlei, Regierung Hitler 1933–1938, part I, 1933/34, volume 1 and 2, Boppard am Rhein 1983.

Reuss, Eberhard: Hitlers Rennschlachten, Berlin 2006.

Rigele, G.: Die Großglockner Hochalpenstraße. Zur Geschichte eines österreichischen Monuments, Vienna 1998.

Rings, Hanspeter: Die Reichautobahn – Monument der Stärke? In: Monika Ryll: Architektur in Mannheim, Mannheim 1994.

Rohde, Heidi: Transportmodernisierung oder Verkehrsbewirtschaftung?, Frankfurt 1999.

Romeyk, Horst: Adenauers Beziehungen zum Rheinischen Provinzialverband und zu staatlichen Behörden, in: Hugo Stehkämper (ed.): Konrad Adenauer. Oberbürgermeister von Köln, Köln 1976, pp. 295–328.

Rose, Mark M (1990): Interstate – Express Highway Politics 1939–1989, Tennessee University Press.

Rossmann, R.: Mythos Bayern, München 2002.

Roth, Ralph (2009): Neue Wege in ein neues Europa: Geschichte und Verkehr im 20. Jahrhundert, edited by Ralf Roth and Karl Schlögel, Frankfurt/Main, 2009.

Rubinstein, James : Making and Selling Cars, New York 2001.

Ruser, Ursula (1980): Die Reichsbahn als Reparationsobjekt, Freiburg.

Ryll, Monika: Architektur in Mannheim, Mannheim 1994.

Sachs, Wolfgang: Die Liebe zum Automobil, Berlin 1984.

Schipper, Frank: Driving Europe – Building Europe on Roads in the Twentieth Century, Amsterdam 2008.

Schivelbusch, Wolfgang: Die Eisenbahnnreise, Frankfurt 2000.

Schivelbusch, Wolfgang (2005): Entfernte Verwandtschaft – Faschismus, Nationalsozialismus, New Deal 1933–1939, Munich 2005.

Schlögel, Karl (2005): Sankt Petersburg – Stadt am Meer, Stadt am Belomor-Kanal, in: Karl Schlögel, Marjampole, München, pp. 141–154.

Schlögel, Karl und Ralf Roth (eds.)(2009): Neue Wege in ein neues Europa: Geschichte und Verkehr im 20. Jahrhundert, Frankfurt/Main, 2009.

Schulz, Günther (ed.): Konrad Adenauer 1917–1933 – Dokumente aus den Kölner Jahren, SH Verlag, Köln 2007.

Schumacher, Angela: „Vor uns die endlosen Strassen, vor uns die lockende Ferne..." – „ Vom Tanken und Rasten auf Entdeckerfahrten durch deutsche Lande", in: Rainer Stommer and Claudia Gabriele Philipp (eds.): Reichsautobahnen – Pyramiden des Dritten Reiches, Marburg 1982, pp. 77–90.

Schütz, Erhard and Eckhard Gruber: Mythos Reichsautobahn, Berlin 1996, 2. edition, Berlin 2000.

Schwarz, Hans-Peter: Adenauer – Der Aufstieg 1876–1952, Stuttgart 1986.

Schwell, Dietrich and Wolfgang Jäger: Verkehrsfreigaben von Autobahnen in den Jahren 1939 bis 1945, in: Strasse und Autobahn, issue 8, 2000, pp. 478–488.

Seely, Bruce E. (1978): Building the American Highway System–Engineers as Policy Makers, Philadelphia 1978.

Seely, Bruce (1998): Der amerikanische Blick auf die deutschen Autobahnen, Werkstatt Geschichte, Hamburg, issue 21, pp. 13–15.

Seely, Bruce (2007): An Overview Essay: Roads in Comparative Perspective, in: Gijs Mom and Laurent Tissot (eds.) (2007): Road History, Neuchatel: Editions Alphil, pp.13–32.

Seidler, Fritz: Fritz Todt, Bublies Verlag, Beltheim-Schnellbach 2000.

Shand, J.D.: The Reichsautobahn: Symbol for the Third Reich, in: Journal of Contemporary History, 1984, Vol. 19, pp. 189–195.

Shannon, Kelly and Marcel Smets: The Landscape of Contemporary Infrastructure, Amsterdam 2010.

Steininger, Benjamin: Raum-Maschine Reichsautobahn, Berlin, Kulturverlag Kadmos 2005.

Stern, Volkhard: Der Autobahn-Schnellverkehr, Freiburg 2007.

Stommer, R. and Claudia Gabriele Philipp (eds.): Reichsautobahnen – Pyramiden des Dritten Reiches, Marburg 1982.

Stommer, R.: Triumph der Technik – Autobahnbrücken zwischen Ingenieuraufgabe und Kulturdenkmal, in: Stommer and Philipp: Reichsautobahnen – Pyramiden des Dritten Reiches, Marburg 1982.

Technische Hochschule Stuttgart (ed.): Paul Bonatz zum Gedenken, Technische Hochschule Stuttgart, 1957.

Thalmann, Frauke (1996): Industrie- und Handelskammern im Ruhrgebiet während der Weimarer Republik, Berlin.

Thewalt, Alexander: Projekte der Autobahnfrühzeit im Ausland – Beispiele aus Italien und den USA, in: W. Wirth: Die Autobahn – Von der Idee zur Wirklichkeit, Cologne 2005, pp. 63–76.

Thompson, Gregory: The Passenger Train in the Motor Age: California Rail and Bus Industries 1910–1941 (Ohio State Univ. Press, 1993.

Tüffers, Bettina: Der Braune Magistrat, Personalstruktur und Machtverhältnisse in der Frankfurter Stadtregierung 1933–1945, Frankfurt 2004 (Studien zur Frankfurter Geschichte Bd. 54).

Urry, John: The Tourist Gaze: Leisure and Travel in Contemporary Societies, London 1990.

Vahrenkamp, Richard (2006): Automobile tourism and political propaganda: constructing the Munich–Salzburg Autobahn 1933–39, Journal of Transportation History, issue 2, 27, pp. 21–38.

Vahrenkamp, Richard (2007): Der Autobahnbau in Hessen bis 1943, Hessisches Wirtschaftsarchiv Darmstadt.

Vahrenkamp, Richard (2008): Lastkraftwagen und Logistik in Deutschland 1900 bis 1955, in: Vierteljahresschrift für Sozial- und Wirtschaftsgeschichte, Bd. 95, 2008, issue 4, pp. 430–455.

Vahrenkamp, Richard (2009): Die Logistische Revolution – Logistik und Güterverkehr in Europa 1950–2000, in: Neue Wege in ein neues Europa: Geschichte und Verkehr im 20. Jahrhundert, edited by Ralf Roth and Karl Schlögel, Frankfurt/Main, 2009, pp. 452–475.

Vahrenkamp, Richard (2010): The Rise of Logistics in Europe 1950–2000, in: European Transport, issue 44.

Vahrenkamp, Richard (2010A): Von Taylor zu Toyota – Rationalisierungsdebatten im 20. Jahrhundert, Köln 2010.

Van Laak, Dirk: Infra-Strukturgeschichte, in: Geschichte und Gesellschaft, 27 (2001), pp. 367–393.

Van Laak, Dirk: Imperiale Infrastruktur: Deutsche Planungen für eine Erschließung Afrikas 1880 bis 1960, Paderborn 2004.

Van der Vleuten, Erik and Arne Kaijser: Networking Europe. Transnational Infrastructures and the Shaping of Europe 1850–2000, Sagamore Beach: Science History Publications, 2006.

Verhoeven, Jennifer: „...hier war einmal Vollkommenheit erreicht." Die Reichsautobahnbrücke von Paul Bonatz bei Limburg a.d. Lahn, in: Denkmalpflege und Kulturgeschichte, issue 2, 2007, pp. 2– 8.

Waitz von Eschen, Friedrich Frhr.: Der nordhessische Braunkohlenbergbau 1578 bis 2003, in: Zeitschrift des Vereins für hessische Geschichte und Landeskunde, 110 (2005), pp. 113–126.

Walsh, Magaret: Making Connections. The Long Distance Bus Industry in the United States, Aldershot 2000

Walter, Martin and Heinz Jordan: Das Herkules-Bergrennen in Kassel, Kassel 2005.

Windisch-Hojnacki, Claudia: Die Reichsautobahn – Konzeption und Bau der RAB, ihre ästhetischen Aspekte, sowie ihre Illustration in Malerei, Literatur, Fotographie und Plastik, Diss. Universität Bonn, 1989.

Winkler, Heinrich: Der entbehrliche Stand – Zur Mittelstandspolitik des „Dritten Reiches“, in:Winkler, Heinrich: Liberalismus und Antiliberalismus: Studien zur politischen Sozialgeschichte des 19. und 20. Jahrhunderts, Göttingen, 1979, pp. 110–144.

Wirth, W. (ed.): Die Autobahn – Von der Idee zur Wirklichkeit, Köln 2005.

Wustrack, Michael: Vom Flughof Rebstock zur Drehscheibe im Weltluftverkehr, in: Ulrich Eisenbach and Gerd Hardach (eds.): Reisebilder aus Hessen, Fremdenverkehr, Tourismus und Kur seit dem 18. Jahrhundert, Darmstadt 2001, pp. 163–176.

Zeller, Thomas (2002): Straße, Bahn, Panorama – Verkehrswege und Landschaftsveränderung in Deutschland 1930 bis 1990. Diss. Munich, Campus Verlag, Frankfurt.

Zeller, Thomas (2007): Driving Germany – The Landscape of the German Autobahn, 1930–1970, New York.

Zeller, Thomas (2009): Der verlangsamte Verkehr: Die Herstellung von Landschaft durch Strassen im 20. Jahrhundert, in: Ralph Roth and Karl Schlögel (eds.): Neue Wege in ein neues Europa, Frankfurt, pp. 361–377.

Zenone, Daniela: Das Automobil im italienischen Futurismus und Faschismus, Forschungsbericht FS II 02-115, 2002, Wissenschaftszentrum Berlin.

Zibell, S.: Jakob Sprenger (1884–1945) – NS-Gauleiter und Reichsstatthalter in Hessen, Darmstadt 1999.

27 Appendix 1: Publications of the Hafraba in 1927

(Source: Libraray of University Braunschweig)

Otzen, Robert: Die Autostraße Hansestädte–Frankfurt–Basel; Schrift Nr. 1; Zweck und Ziele des Vereins "Hafraba" – Verlag der Göhmannschen Buchdruckerei, Hannover 1927

Tagungsbericht der 1. Verwaltungsrat-Sitzung am Donnerstag, 10. Febr. 1927, in den Räumen der Frankfurter Gesellschaft für Handel, Industrie und Wissenschaft in Frankfurt a/M, Schrift Nr. 2 , Hellerdruck Frankfurt 1927

Vorträge gehalten anlässlich der 1. Verwaltungsrat-Sitzung am 10. Februar 1927 in den Räumen der Frankfurter Gesellschaft für Handel, Industrie und Wissenschaft in Frankfurt a/M/, von Piero Puricelli und Robert Otzen, Schrift Nr. 3 – Hellerdruck Frankfurt 1927

Otzen, Robert: Die Autostraße Hansestädte–Frankfurt–Basel, Erläuterungsbericht zum Entwurf der Autostraße, Schrift Nr. 4 – Hellerdruck Frankfurt 1927

1. Sitzung des Technischen Ausschusses am Dienstag, den 5. April 1927 in Frankfurt am Main, Schrift Nr. 5 – Hellerdruck Frankfurt 1927

1. Sitzung des Verkehrswirtschaftlichen Ausschusses am Montag, den 16. Mai 1927, in den Räumen der "Hafraba" Frankfurt am Main Savignystr. 25, Schrift Nr. 7 – Hellerdruck Frankfurt 1927

2. Sitzung des Technischen Ausschusses am Dienstag, den 24. Mai 1927, im Hotel Disch in Köln a.Rh., Schrift Nr. 7 – Hellerdruck Frankfurt 1927

2. Sitzung des Verwaltungsrates am Mittwoch, den 25. Mai 1927, im Sitzungssaal des Hotel Disch in Köln am Rhein, Schrift Nr. 8 – Hellerdruck Frankfurt 1927

2. Sitzung des Verkehrswirtschaftlichen Ausschusses am Samstag, den 29. Oktober 1927, im Turmsaale des Rathauses zu Mannheim, Schrift Nr. 9 – Hellerdruck Frankfurt 1927

1. Sitzung des Finanzierungs-Ausschusses am Donnerstag, den 1. Dezember 1927, im Hotel Ernst-August zu Hannover, Schrift Nr. 10 – Hellerdruck Frankfurt 1927

3. Sitzung des Verwaltungsrats am Donnerstag, den 1. Dezember 1927, im Hodlersaal des Neuen Rathauses zu Hannover, Schrift Nr. 11 – Hellerdruck Frankfurt 1927

28 Appendix 2: Studies on the German autobahn project

K. Kaftan: Der Kampf um die Autobahn, Berlin 1955

*Paul Hafe*n: Das Schrifttum über die deutschen Autobahnen, Bonn 1956.

Karl-Heinz Ludwig: Technik und Ingenieure in Dritten Reich, Königstein 1974

Karl Lärmer: Autobahnbau in Deutschland 1933 bis 1945, Berlin 1975

H. Henning: Kraftfahrzeugindustrie und Autobahnbau in der Wirtschaftspolitik der Nationalsozialisten 1933 bis 1936, in: Vierteljahresschrift für Sozial- und Wirtschaftsgeschichte, Bd. 65, 1978, pp. 217–242

R. Stommer (Hersg.): Reichsautobahnen – Pyramiden des Dritten Reiches, Marburg 1982

Wolfgang Sachs: Die Liebe zum Automobil, Berlin 1984.

J. D. Shand: The Reichsautobahn: Symbol for the Third Reich, in: Journal of Contemporary History, 1984, Vol. 19, p. 189–195

Anette Nietfeld: Reichsautobahn und Landschaftspflege, Werkstattberichte des Instituts für Management und Umweltplanung, herausgegeben von *Arnim Bechmann*, Nr. 14, 1985, TU Berlin.

Claudia Windisch-Hojnacki: Die Reichsautobahn – Konzeption und Bau der RAB, ihre ästhetischen Aspekte, sowie ihre Illustration in Malerei, Literatur, Fotographie und Plastik, Diss. Universität Bonn, 1989

M. Kornrumpf: HAFRABA e.V. – Deutsche Autobahn-Planung 1926–1934, Bonn 1990. A study without clearing the sources and with admiration of the NS-movement.

Christopher Kopper: Modernität oder Scheinmodernität nationalsozialistischer Herrschaft – Das Beispiel der Verkehrspolitik, in: *Christian Jansen, Lutz Niethammer und Bernd Weisbrod* (Hersg.): Von der Aufgabe der Freiheit, Festschrift für Hans Mommsen, Berlin 1995, pp. 399–411

E. Schütz und E. Gruber: Mythos Reichsautobahn, Berlin 1996, 2. edition, Berlin 2000

Franz Seidler: Fritz Todt, Bublies Verlag, Beltheim-Schnellbach 2000.

Dietrich Schwell und *Wolfgang Jäger*: Verkehrsfreigaben von Autobahnen in den Jahren 1939 bis 1945, in: Strasse und Autobahn, issue 8, 2000, pp. 478–488.

Thomas Zeller, Straße, Bahn, Panorama – Verkehrswege und Landschaftsveränderung in Deutschland 1930 bis 1990. Diss. Munich, Campus Verlag, Frankfurt 2002

Steininger, Benjamin: Raum-Maschine Reichsautobahn, Berlin, Kulturverlag Kadmos 2005.

Thomas Zeller: Driving Germany – The Landscape von the German Autobahn 1930–1979, New York, 2007.

Vahrenkamp, Richard: Autobahnbau in Hessen bis 1943, in: Hessisches Wirtschaftsarchiv, Darmstadt, Wirtschaftsarchiv, 2007

29 Abbrivations:

ADAC	Allgemeiner Automobilclub Deutschland
ADM	Archiv Deutsches Museum, München, Seifert Nachlass
ADR	Akten der Reichskanzlei
AHR	Archiv Hessicher Rundfunkt, press cuttings Autobahn
BA	Bundesarchiv Berlin
DIHT	Deutscher Industrie- und Handelstag
DM	Deutsche Mark. In the 1950s were 4,20 DM equal to 1 US Dollar.
GStA PK	Geheimes Staatsarchiv Preußischer Kulturbesitz
GEZUVOR	Gesellschaft zur Vorbereitung des Reichsautobahnbaus e.V.
HAFRABA	Verein zur Vorbereitung der Autostraße Hansestädte–Frankfurt–Basel
HAStK	Historisches Archiv Köln
HHSTAW	Hessisches Hauptstaatsarchiv Wiesbaden
HP	Horse power
IHK	Industrie- und Handelskammer
ISG	Institut für Stadtgeschichte Frankfurt a.M.
Jg.	Jahrgang
Kfz	Kraftfahrzeug
LKW	Lastkraftwagen
Mio. :	Million
NS	nationalsozialistisch
NSKK	Nationalsozialitisches Kraftfahrkorps
OT	Organisation Todt
PKW	Personenkraftwagen
RGB	Reichsgesetzblatt
RM	Reichsmark
SA	Sturmabteilung
SM	Stadtarchiv München, ZA Autobahnen
SS	Sturmstaffel
UBK:	Universitätsbibliothek Kassel, Hessischer Lesesaal
WWI	World War I

Editorial Note

Parts of this books already appeard in the following journals and books:

Vahrenkamp, Richard (2006): Automobile tourism and political propaganda: constructing the Munich–Salzburg Autobahn 1933–39, Journal of Transportation History, Vol. 27, issue 2, 27, pp. 21–38. I thank Dr. Matthew Frost from Manchester University Press for granting the copyright.

Vahrenkamp, Richard (2006): Die Chiemsee-Autobahn. Planungsgeschichte und Bau der Autobahn München–Salzburg 1933–1938; in: Oberbayerisches Archiv, vol. 130, Verlag des Historischen Vereins von Oberbayern, München 2006, pp. 385–416.

Vahrenkamp, Richard (2007): Der Autobahnbau in Hessen bis 1943, Hessisches Wirtschaftsarchiv Darmstadt. I thank Dr. Eisenbach, the head of the Hessian Economic Archive in Darmstadt, for granting the copyright.

Vahrenkamp, Richard (2007): Planning and Constructing the Autobahn Network in Germany during the Nazi-Period: Styles and Strategies, in: Gijs Mom and Laurent Tissot (eds.): Road History – Planning, Building and Use, Collection History of Transport, Tourism and Travel, Editions Alphil, Neuchâtel 2007, pp. 87–98.

Vahrenkamp, Richard (2009): Verkehrspolitik und Autobahnbau in Nordhessen 1925 bis 1955, in: Zeitschrift des Vereins für hessische Geschichte und Landeskunde, vol. 114, 2009, pp. 235–269.

Vahrenkamp, Richard (2010): Motorization and Autobahn Projects in Germany in the Interbellum, in: Transportes, Servicios y Telecomunicaciones, June 2010.

EINZELSCHRIFTEN

Wolfgang Chr. Fischer (Ed.)
German Hyperinflation 1922/23 – A Law and Economics Approach
Lohmar – Köln 2010 ◆ 220 S. ◆ € 59,- (D) ◆ Hardcover ◆ ISBN 978-3-89936-931-1

Thomas Eschenbach
Zum Budgetdefizitkriterium des Stabilitäts- und Wachstumspaktes
Lohmar – Köln 2010 ◆ 200 S. ◆ € 49,- (D) ◆ ISBN 978-3-89936-932-8

Marco Dreher
Unternehmenswertorientiertes Beteiligungscontrolling – Aufgabenspezifische Fundierung auf Basis entscheidungs- und kapitalmarktorientierter Konzepte der Unternehmensbewertung
Lohmar – Köln 2010 ◆ 556 S. ◆ € 76,- (D) ◆ ISBN 978-3-89936-933-5

Laura Gerardy, Johannes Vogel, Leif Chr. Steinbrinker und Sybille Falke
Zukunftsmodell Ärztenetzwerke – Eine empirische Untersuchung der Auswirkungen auf die Healthcare Industrie
Lohmar – Köln 2010 ◆ 120 S. ◆ € 42,- (D) ◆ ISBN 978-3-89936-934-2

Marco Vietor
Studienfinanzierung durch Humankapitalverträge – Eine Marktanalyse
Lohmar – Köln 2010 ◆ 200 S. ◆ € 49,- (D) ◆ ISBN 978-3-89936-935-9

Henning Struck
Mandatsunfähigkeit von Aufsichtsratsmitgliedern einer deutschen Aktiengesellschaft
Lohmar – Köln 2010 ◆ 252 S. ◆ € 57,- (D) ◆ ISBN 978-3-89936-937-3

Richard Vahrenkamp
The German Autobahn 1920–1945 – Hafraba Visions and Mega Projects
Lohmar – Köln 2010 ◆ 280 S. ◆ € 58,- (D) ◆ ISBN 978-3-89936-940-3